Life through the Ages

LIFE OF THE PAST

James O. Farlow, Editor

LIFE *through the* AGES

COMMEMORATIVE EDITION

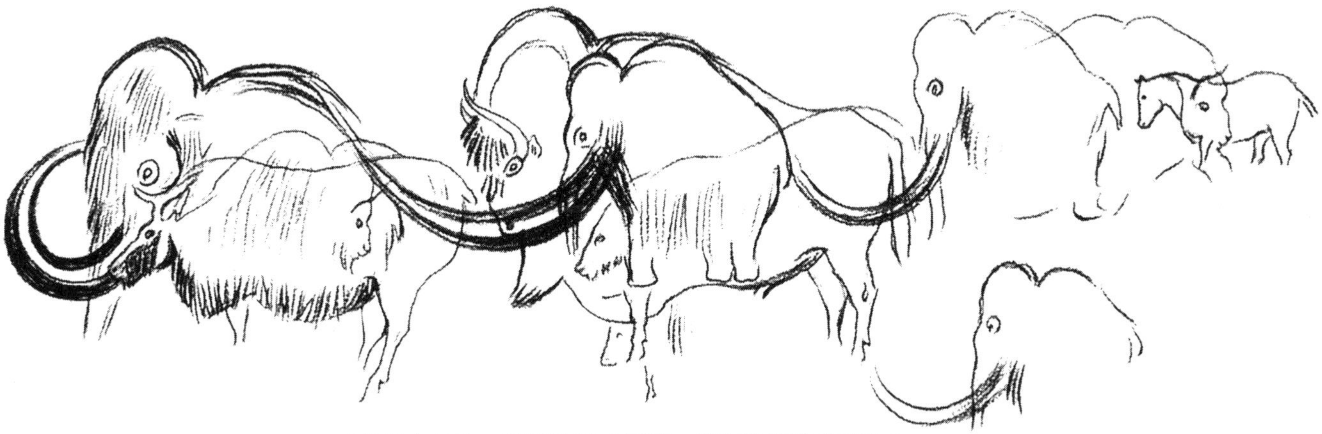

FRIEZE. SHOWING MAMMOTHS. HORSES. BISON AND REINDEER. CAVERN OF THE FONT-DE-GAUME · LES-EYZIES · DORDOGNE · FRANCE

CHARLES R. KNIGHT

New Foreword by

Stephen Jay Gould

Introduction by

Philip J. Currie

INDIANA UNIVERSITY PRESS • BLOOMINGTON AND INDIANAPOLIS

This book is a publication of

Indiana University Press
601 North Morton Street
Bloomington, Indiana 47404-3797 USA

http://iupress.indiana.edu

Telephone orders 800-842-6796
Fax orders 812-855-7931
Orders by e-mail iuporder@indiana.edu

The paper used in this publication meets the minimum requirements of American National Standard for InformationSciences—Permanence of Paper for Printed Library Materials, ANSI Z39.48-1984.

Manufactured in China

Library of Congress Cataloging-in-Publication Data

Knight, Charles Robert, 1874–1953.
Life through the ages / Charles R. Knight ; new foreword by Stephen Jay Gould ; introduction by Philip J. Currie.— A new ed.
p. cm.
Originally published: New York : Alfred A. Knopf, Inc., 1946.
Includes bibliographical references.
ISBN 0-253-33928-6 (cl : alk. paper)
1. Paleontology. I. Title.
QE763 .K5 2001
560—dc21

2001016581

1 2 3 4 5 06 05 04 03 02 01

Contents

Foreword: Life through Our Ages

Stephen Jay Gould

ALEXANDER AGASSIZ PROFESSOR OF ZOOLOGY

HARVARD UNIVERSITY

Like so many unprivileged urban kids who became paleontologists, I received my inspiration from dinosaur skeletons at museums (then without entrance fees) and books on the history of life from public libraries. New York's Museum of Natural History was my church, and since churches feature saints and doxologies, I well remember the statue of Henry Fairfield Osborn, the powerful and patrician president of the museum, America's most famous vertebrate paleontologist, and Charles R. Knight's most important fan and patron. I memorized the inscription below the statue, not by conscious intent, but rather by constant reading: "For him the dry bones came to life and giant forms of ages past rejoined the pageant of the living."

That text might now be judged a bit stilted in its post-Victorian cadences, but the content cannot be gainsaid. However, these true and noble words are, on this statue, associated with the wrong person. For they belong, by any standard of justice or accuracy, to Charles R. Knight, not to Henry Fairfield Osborn. I do not mean to denigrate Osborn in any way. He was a great scientist (a tad on the conservative side, perhaps), and an equally great administrator, with ties of blood and wealth to all the power in New York City (and elsewhere) that could nurture and finance his beloved museum. He also appreciated the unique genius and combination of Knight's artistic talents and anatomical knowledge, and he established Knight, virtually canonized him, as the preeminent authority for the reconstruction of ancient organisms.

Authors of text tend to persist in memory as the creators of the past, for our cultural traditions have become primarily verbal. But primates are, preeminently, visual animals—and no cultural invention, not even Gutenberg's press, so often judged as the greatest technological achievement of the second millennium, can extinguish our ineradicable preferences for visual representations. Confucius didn't really say the famous words conventionally attributed to him: "One picture is worth a thousand words." But who would deny the evident truth of this profound, if anonymous, aphorism?

The main point of this foreword, and the main rationale (one might even say necessity) for reprinting this best of Knight's books, can thus be simply stated: If one seeks the name of the person most responsible for our usual sense, our everyday "feel," of the nature, status, beauty, strangeness, and fascination of prehistoric life, only one person can possibly claim the title—and his credentials simply cannot be denied. That person is not Osborn or any other paleontologist, and not Darwin or any other biological theorist or naturalist. Indeed, the name of that person remains largely unknown or little regarded, even by many people with a keen interest in the subject, because we honor writers of text, while makers of images tend to remain anonymous. That man is Charles R. Knight. He virtually owned the genre of accurate artistic reconstruction of prehistoric life, from his first works of the 1890s, to his acme in the 1920s, and continuing to his death in the 1950s. Essentially all the standard twentieth-century pictures are his—all the murals in major museums, all the paintings in "coffee table" books—and he executed these works with unparalleled accuracy and artistic merit.

Knight created the standard imagery of dinosaurs before a host of novel discoveries, and a new band of young and gifted artists, restructured our concepts and our pictures, but only in the last generation of the twentieth century. The canonical appearance of ancient life belongs to Charles R. Knight, and his errors were only the accurate renderings of scientific waystations in the endless saga of increasing knowledge. Those memorized words of my youth belong to Charles R. Knight—for he, more than any other person, revivified the dry bones and placed them, as animals, within the pageant of vitality. Great art, as Plato stated, need not falsify reality or make nature "artificial" (in the word's pejorative sense). Great art improves our understanding of nature's reality.

Knight, a New Yorker by birth and upbringing, trained to become a commercial artist and received his first steady employment with the church-decorating firm of J. and R. Lamb in the 1890s. But he loved natural history most of all, and spent several mornings each week sketching animals at the Central Park Zoo. He moved from being a specialist on plants and animals for his firm's stained-glass windows to a full career in zoological art, eventually building a special reputation for the reconstruction of extinct organisms. Osborn's patronage, and Knight's resulting work for New York's Museum of Natural History, gave him a permanent base and semiofficial status as the preeminent iconographer of prehistoric life.

Life through the Ages, published in 1946, and my favorite of Knight's works, best displays both his strengths and the weaknesses of professional understanding more than half a century ago—all shown with double clarity, because Knight wrote his own text to accompany his masterful drawings. He also treated (albeit with far greater emphasis on creatures regarded as "higher" or "more interesting" by his human readers) the entire history of multicellular life, from Cambrian creatures of the Burgess Shale to battles between cave men and cave bears. I think that we can identify two major sources of general difference (as opposed to particular details of change wrought by more knowledge of particular fossils) between his concepts and ours—and I don't think that we can be deemed smug or arrogant in suspecting the higher accuracy of our current views, based largely upon empirical discoveries and theoretical breakthroughs since Knight's time, and perforce unavailable to him. (Knight was a great artist, but not a conceptual innovator. He conceived his task as rendering the best professional opinion of his day in iconography that would both focus thought and stimulate visceral fascination.)

First, Knight followed the progressivist, or even more strongly triumphalist, view that construed the history of life as a rising struggle through blood, toil, sweat, and tears toward a human pinnacle. This conception led him, sometimes subtly and unconsciously, sometimes quite overtly, to equate old with dumb, lumbering, and inefficient, and geologically more recent with smart and lithe. Such a view, we would now argue, does not adequately honor the differences and fascinations of our more distant forebears. Dinosaurs are bound to get shorter shrift than they deserve if their bodies and environments truly merit Knight's description herein: "Nature had apparently grown weary of the great scaly cold-blooded monsters. They had been in existence too long, for they were stupid, unadaptable, and unprogressive. And so the world was to grow away from these slow-moving dunces, and little warm-blooded beings, furry, alert and aggressive, were to supersede them. . . . It must have been a most depressing world as we think about it now, with huge, bizarre and ungainly shapes rising and subsiding in the landscape, the earth covered with harsh and brittle scrub under tall palmettos."

Second, Knight followed the limited knowledge

and social conventions of a more martial age that did view Darwinism as an overt and literal struggle for existence by murder and mayhem (the restricted notion that T. H. Huxley criticized as the "gladiatorial theory of existence")—and not, in Darwin's actual formulation and broader intent, as competition for differential reproductive success by a full range of various strategies most appropriate for particular circumstances (including cooperation as well as combat). Knight therefore depicts all interactions among species as threats, chases, and kills, and never as cooperation in feeding, or even as parental nurturing. To me, the most silently revealing of all Knight's drawings (one picture is worth a thousand words, after all) combines both these conceptual archaicisms by placing *Archaeopteryx,* the first bird, higher on a branch and engaged in aggressive posturing with a spatially (and, in Knight's view, also taxonomically) lower flying reptile. In fact, if any such struggle ever occurred, I'd make a substantial wager on the pterosaur.

But to criticize Knight for not knowing what later generations would learn would be as ridiculous as lambasting Leonardo because, living before Copernicus, he still viewed the cosmos as geocentric. We need Knight's work in press today and forever, because he was history's greatest artist of past life—the Ruth of his age (and who would cease to honor the Bambino just because a guy named McGwire hit ten more homers in a single season some seventy years later?). We need to know Knight because he, more than any other person in history, created our conventional imagery of prehistoric life. And we need to discuss Knight's errors as well as his triumphs, if only to remind us that many of our current certainties will seem just as false, but equally instructive as mirrors of their time, to the next generation of scribes and artists, the guardians and creators of our growing knowledge of the rich and wondrous history of this ancient planet.

Introduction

Philip J. Currie
CURATOR OF DINOSAURS
ROYAL TYRRELL MUSEUM OF PALAEONTOLOGY
DRUMHELLER, ALBERTA, CANADA

It was Roy Chapman Andrews who convinced me as an eleven-year-old to make a career out of hunting dinosaurs. He did not do this directly, of course, but his book *All about Dinosaurs* had such an impact on me that I found myself scrounging up everything I could get on dinosaurs. One of my first prizes in this quest was an Audubon sticker book about prehistoric animals. I often sat and admired the stickers, which were reproductions of the stunningly beautiful color paintings that Charles Knight had done for the Field Museum of Natural History. Such was the power of those images that they worked their way into my dreams, and started to shape my understanding of how dinosaurs interacted with each other and their environments. A year later, a family vacation took us to the hallowed halls of the Field Museum in Chicago, where I stood in open-mouthed wonder in front of the giant murals that had been reproduced in my sticker book. To this day, those paintings are burned deeply into my memory, and Charles Knight is the director of my imaginary Museum of Dinosaur Art!

Charles Knight's influence stretched much further than the minds of eleven-year-olds. Even a quick survey of dinosaur art done before the 1970s reveals that he was the main inspiration for most of the book illustrators and museum artists depicting dinosaurs and other prehistoric beasts. If copying is considered to be a form of flattery, he was clearly well admired by artists depicting prehistoric animals for novels, cartoons, comic books, and a host of other fictional spin-offs.

The artwork created by Knight was so powerful that his images continued to leave a strong impression on both public and scientific minds long after his death. Most people were loath to see their childhood images of prehistory changed when a wealth of new information became available in the 1970s. Ever so slowly at first, the subconscious, inertial reactions were overcome by new and better evidence that came with the development of the "Dinosaur Renaissance" (Bakker 1975). Once the changes started, they opened up the floodgates for a whole new generation of artists producing completely original dinosaur art (Crumly 2000). The influence of Charles Knight may be less obvious in the work of these new artists, but there can be no denying that they are carrying on traditions with their foundations in Knight's work.

There are many reasons for Knight's long-standing popularity. His drawings and paintings are aesthetically pleasing. But more important, he was able to instill his considerable knowledge of anatomy (of both modern and fossil animals), ecology, and animal behavior into his artwork to create realistic, believable, lifelike scenes.

Knight learned much through his long association with professional paleontologists such as Henry Fairfield Osborn of the American Museum of Natural History. Although he did the best that he could in terms of accuracy, our ideas about prehistoric animals are constantly evolving and changing as new information is acquired from the fossil record. This includes the discovery of more complete fossils, and the acquisition of new or better information through further study of existing specimens and fossil sites. Therefore, even though Knight's illustrations were as accurate as they could have been at the time he did them, changes in our knowledge base have dated some of his illustrations.

Life through the Ages was both illustrated and written by Charles Knight, and was published in 1946. In many ways it was a remarkable book, far ahead of its time. It is only recently that it has become common for illustrators to write their own text for popular books (Paul 1988; Dixon 1993). Knight's text not only provides the reader with information, but also gives some insight into his artwork, explaining the decisions he made in composing the illustrations. Knight was a proponent of using the present as a key to understanding the past, and the book is a precocious combination of prehistoric and recent biology. He was able to make his extinct animals come to life, and after more than half a century, many of the illustrations are still scientifically sound. This is particularly true for those animals with living relatives, such as members of the cat, dog, elephant, and horse families. Restorations of other animals, however, would be done differently by modern artists, who benefit from the changes that have occurred in our knowledge over the last half-century. Similarly, there are many ways that Knight would have done things differently if he were still alive to see how our knowledge of prehistoric animals has changed. Some of the most profound changes in our understanding of extinct life forms have taken place with dinosaurs. As dinosaur research is the field that I know and love the best, I will use them as an example of how Knight might have done things differently if he were alive today.

Knight starts his book not with dinosaurs, but with the Burgess Shale fauna from British Columbia. In recent years, this and similar soft-bodied Cambrian faunas from other parts of the world have attracted ever-increasing amounts of attention. As Knight stated, "there is still a good deal of dispute as to their actual affinities."

Sauropods were seen by Knight and his contemporaries as "water-living dinosaurs," each of which "spent practically all of its time in the water, feeding on soft vegetation which it bit off with its small finger-like teeth," and which were too heavy to hold up their own weight on land. These incredible animals, which include the longest (*Diplodocus, Seismosaurus*), heaviest (*Argentinosaurus*), and tallest (*Brachiosaurus, Ultrasauros*) species of land-dwelling (terrestrial) animals known, were one of the most successful lineages of dinosaurs. They lived almost worldwide, and their history spanned virtually the entire Jurassic and Cretaceous. It had long been thought that sauropods spent most of their time in the water, which would have supported their enormous body weights. This may have been true for some species, but many lines of evidence (including studies of their distribution and their trackways) have shown that sauropods were capable of walking efficiently on land. In fact, some may have sought their food high in the trees, browsing much like modern giraffes. Sauropod teeth are relatively small and weak compared with overall body size. However, it was not necessary for them to be strong, because they functioned only to comb off branches, and had no role in chewing. The vegetation was swallowed without any further processing in the mouth. The "chewing" was done by gastroliths, stones swallowed and held in a muscular sac in the digestive system. Many modern birds swallow pebbles, which are used for much the same function. Stones are held in the muscular gizzard to grind up seeds and other vegetation. The advantage for sauropods to have used the same system as in birds is that they were able to keep the weight of their teeth to a minimum. This was critical for a sauropod because of the enormous length of the neck, which can be as long as the rest of the body. It would have been mechanically inefficient for a sauropod to have such a long neck if the head were too heavy. Small teeth meant that the skull could be kept small, and having relatively weak jaws that were not involved in chewing meant that the jaw muscles could also be kept small. Selection may even have favored the development of relatively small brains in sauropods as a means of reducing weight. Although the teeth of sauropods may have been relatively weak, they were still more than adequate to ingest a variety of plants, and these did not have to be soft water plants, as was once thought. In fact, sauropod teeth often show extreme wear, and the presence of several generations of germ teeth beneath each functional tooth suggests that teeth were replaced frequently.

Although their feeding strategies may still be a controversial topic, very few modern depictions show sauropods floating in the water. This was once the common belief, and was based on the supposition that sauropods were so large and heavy that they needed to spend most of their lives in water to buoy up their great weights. Evidence—including dinosaur trackway sites, the analysis of skeletal structures, and the recovery of sauropods from regions that were relatively dry when the animals lived there—now suggests that most sauropods were probably more comfortable on land than they were in the water. In spite of his belief that sauropods lived mainly in the water, Knight seemed to hedge his bets in his illustrations, because he often showed them both on land and in the water.

An interesting aside in the text is that sauropods "probably had living young." This idea has most recently been championed by Bakker (1986), whose evidence seems reasonable. However, sauropod eggs are known from many parts of the world (Carpenter 1999), some with embryos inside (Chiappe 1998). The fact that sauropods are known to lay eggs does not, however, eliminate the possibility that some species may have given live birth. This is an attribute that can develop in select species within animal lineages that normally lay eggs, including insects, fish, amphibians, and reptiles.

Knight dealt with *Tyrannosaurus rex* as the final dinosaur because it was "the grandest and fiercest dinosaur"—"the last tremendous example of a mighty race." *T. rex,* probably the best-known dinosaur, may well have been "an enormous eating machine with an insatiable appetite," but among dinosaurs it was reasonably well endowed with a larger than average brain. By the end of the Cretaceous, there was a tendency toward increased brain size for almost all dinosaurs, and it was necessary for the King Tyrant to keep pace with its prey (ceratopsians and hadrosaurs). As Knight pointed out, *T. rex* was the culmination of millions of years of theropod evolution, and it was the most highly specialized large carnivore that we are aware of. In recent years, larger theropods have been discovered in Argentina (Coria and Salgado 1995), but these carcharodontosaurids had smaller brains, lacked binocular vision, and were stouter, slower animals with relatively short lower leg proportions. No other animal is known with the biting power of *T. rex,* which had enough strength in its jaw muscles to drive the disproportionately long, heavy teeth right into or through the bones of its victims (Erickson and Olson 1996).

The illustration of the two tyrannosaurs fighting would have been done differently if Knight had had access to the information available today. First and foremost, the lizard-like tails would not have been on the ground. They were reconstructed too long in the original American Museum skeletal mounts (complete tails of *Tyrannosaurus* were unknown until recently). Although they may have touched the ground periodically, they were balancing organs that served the animal best by being held in a more horizontal position off the ground. To conserve energy, overlapping parts of the vertebrae and the chevrons beneath them helped to prevent the tail from drooping. The heavily scaled skin of the tyrannosaur drawing is probably incorrect as well. At least three tyrannosaur specimens have been found in recent years with patches of preserved skin. Although the skin is not known from the entire surface of the body, those patches suggest that tyrannosaurid hides had lightly pebbled surfaces. Tyrannosaurs appear to have been social animals that hunted in packs (Currie 2000), and apparently engaged in aggressive behavior with each other (Tanke and Currie 2000). This seems to have consisted mostly of face biting, which may have been related to sexual interaction as in modern animals such as sea otters.

The conventional wisdom at the time that Knight wrote his book was that *T. rex* was descended through a lineage of theropods that became progressively larger from Triassic through Jurassic to Cretaceous times. In recent years, evidence has emerged to suggest that *Allosaurus* and tyrannosaurs are not particularly close in relationship, and that in fact *T. rex* has closer ties with animals such as *Ornithomimus* and *Velociraptor.* Similarities between *Tyrannosaurus* and *Allosaurus* are attributed to mechanical problems related to large size. As large animals, they both had to reinforce and modify their skeletons in similar ways, in spite of the fact that they had different family backgrounds. Curiously, the fact that *Tyrannosaurus* and its kin (*Albertosaurus, Gorgosaurus, Tarbosaurus,* etc.) are overgrown coelurosaurs (a name applied to a suite of generally small, meat-eating dinosaurs) was recognized by Frederick von Huene in 1926 when he established the basic division of flesh-eating dinosaurs into two groups—the diminutive coelurosaurs (including

Compsognathus, Ornithomimus, Troodon, and *Velociraptor*) and the giant carnosaurs (*Acrocanthosaurus, Allosaurus, Megalosaurus*).

Stegosaurus is an animal with a reputation for having a small brain relative to its body size. Knight referred to it as the "stupidest member of a very moronic family." Stegosaurs are now known from Africa, Asia, Europe, and North America, and had a long history within the Jurassic and Cretaceous (Galton 1997). Like ankylosaurs and sauropods, stegosaurs did have relatively small brains, but they still managed to be relatively successful. Fortunately for them, large brains are not essential for survival, and they survived for long enough to spread across the globe. Furthermore, there has always been some doubt concerning the relationship between brain size and "intelligence."

Styracosaurus may have been a formidable animal "with his nose horn and sharp turtlelike beak," but Knight was incorrect when he suggested that it sometimes had to fend off a hungry *Tyrannosaurus.* The two animals both lived in western North America, but *Styracosaurus* was extinct at least five million years before the first *Tyrannosaurus rex* appeared. Nevertheless, *Gorgosaurus* and *Daspletosaurus,* earlier members of the family Tyrannosauridae, probably did hunt and kill *Styracosaurus* whenever they could. "*Styracosaurus* was really a harmless old fellow, a vegetable feeder, armed only for defense against marauding flesh eaters." Knight, like most of his contemporaries, believed that the frills over the necks of horned dinosaurs were to protect them from tyrannosaurs. In recent years, a lot of work has been done on ceratopsian bonebeds in Alberta and Montana. These seem to represent mass death assemblages of herds of *Centrosaurus, Einiosaurus, Pachyrhinosaurus,* and *Styracosaurus.* The spiked frill may have helped protect the neck of an adult *Styracosaurus* to an extent, but the frills of young individuals were short and lacked spikes. There is even some indication that the frills and spikes differed between adult males and females. Given the fact that tyrannosaur jaws were capable of biting through even thicker bones than the neck frill of *Styracosaurus,* and the fact that only the adults had well-developed frill spikes, it seems more likely that the frills developed into such fantastic forms in horned dinosaurs so that they could visually recognize other members of their own species. Dinosaurs were apparently highly visual animals, and seem to have used frills, spikes, crests, horns, and other visual cues to attract potential mates, and perhaps even to establish dominance within social groups. This was probably important at the time *Styracosaurus* lived (about 75 million years ago), because horned dinosaurs were at their peak of diversity, and many closely related species lived in the same region. It would have been very easy to make mistakes that would have resulted in unproductive matings. The extreme differences in the frills and horns of ceratopsian adults emphasized species differences, making it more likely that adults would recognize and mate with only members of their own species.

"If ever a treasure was unearthed among all the various remains that have come down to us through the years, surely it is *Archaeopteryx,* the answer to the fossil-hunters' prayer." In spite of a virtual "gold rush" of new early bird fossils from northeastern China, including approximately a thousand specimens of the early Cretaceous *Confuciusornis, Archaeopteryx* is as critical to our understanding of bird origins today as it was when Knight wrote this statement in 1946.

In rereading *Life through the Ages,* I was surprised to read that Knight believed that *Archaeopteryx* was a link between theropod dinosaurs and modern birds. He specifically states that "we shall see pictures of dinosaurs and the birdlike forms which we think may have descended from them." It becomes evident later in the book that the "birdlike forms" include *Archaeopteryx* and several toothed birds from the Cretaceous. This statement, as much as any, shows his belief that dinosaurs were involved in the ancestry of birds. This was not a new idea in 1946. The hypothesis had been worked out by Huxley in 1868 and was widely accepted for the rest of that century. In 1916, Heilmann published a thorough analysis of the evidence for bird origins in his native Danish. Although his work did not have much impact at first, this changed dramatically when *The Origin of Birds* appeared in English (Heilmann 1927). He felt that *Archaeopteryx* was anatomically closer to coelurosaurian dinosaurs such as *Compsognathus* than to any other known animals. This demonstrated to him that birds and dinosaurs were closely related. However, several features suggested to him that dinosaurs were not directly ancestral to birds. For example, the clavicle or collar bone had never been reported in a dinosaur. Birds have clavicles, just as most fish, amphibians, reptiles, and mammals do. If dinosaurs lacked clavicles, this was a specialization that indicated they could not be bird ancestors. After all, if they had lost their clavicles, how could their descendants get them back? Because birds retain the clavicle, then, their ancestor must have still had one. Heilmann felt that if dinosaurs had not given rise to birds, then their ancestors, at that time known as thecodonts, must have. Thecodonts at least still had their clavicles. This was the prevailing theory until the 1970s, when several opposing ideas arose. Regardless of Heilmann's view, Knight clearly thought there was a relationship between birds and dinosaurs. He described *Allosaurus* as "a plucked turkey with a long tail and little front legs with claws instead of wings." "Indeed, as we study them closely, their birdlike affinities become more and more apparent as we shall later see in the *Archaeopteryx* (the bird reptile), a sort of little dinosaur with wings instead of front legs, and feathers, which further enhanced the birdlike appearance."

Between the publication of the Danish and English editions of *The Origin of Birds,* the first wishbone was found in the dinosaur *Oviraptor.* Unfortunately, it was misidentified as a different bone (an interclavicle) for another fifty years. New discoveries and careful reexamination of previously collected specimens have now shown that the wishbone is present in many lineages of theropod families, including tyrannosaurids. The mass of evidence gathered since Knight's book was published strongly favors the hypothesis that birds are the direct descendants of theropod dinosaurs. In fact, the distinction between the two has become blurry with the discovery of at least five species of "feathered" dinosaurs since 1996. Under a modern biological or paleontological classification, birds are even considered to be a subgroup of theropod dinosaurs. In that sense, birds are dinosaurs and dinosaurs are not extinct.

One of Knight's favorite subjects for illustration, birds are compared and contrasted in a suite of sketches. From ancient to modern, he depicted the toothed (yet highly specialized) Hesperornis, the somewhat comparable modern loon, and the gigantic *Phorusrhacus* (*Phororhacos*), looking like some Mark II *T. rex.* As he points out in the text, "our actual knowledge of fossil bird forms is scant, for the fragile skeletons of these flying creatures were easily destroyed." In recent years, fossil sites around the world have produced a wealth of new specimens of early birds. Several of these animals, including more than a thousand specimens of *Confuciusornis* from the Early Cretaceous of China, have feathers for display purposes, including crests on the head and a pair of long tail feathers reminiscent of those of a modern lyretail. Even earlier forms, such as the feathered dinosaur *Caudipteryx* from the same beds, also appear to have developed long feathers on the arms and tail for the purpose of display (Currie 1998). This makes Knight's statement that "there were no beautiful feathered beings in our early world" all the more curious.

Knight showed *Archaeopteryx* sharing a perch with *Rhamphorhynchus,* a pterosaur from the same time and place. It was the old facing the new—the beginning of the end for the pterosaurs. Although flying reptiles were very diverse and successful in colonizing the air, the new feathered fliers were ultimately more efficient, better fliers that slowly replaced their distant cousins. By the end of the Cretaceous, birds had won out in most niches, although they left the role of giant gliders to flying reptiles such as Pteranodon (shown in the next picture with mosasaurs).

Knight shared with many of his contemporaries the idea that dinosaurs were "slow-moving dunces" that were "unadaptable, and unprogressive." None of these descriptive terms are correct for the Dinosauria overall, but they might be applied to certain species in the same way that mammalian sloths are slow, and dodo birds were unadaptable. Although dinosaurs were generally not large-brained animals, some species had brain sizes that were within the lower limits for modern mammals and birds. *Troodon,* from the Late Cretaceous of North America, had a brain six times larger than that of a crocodile of equivalent body weight. Regardless of brain size, dinosaurs seem to have been extremely well adapted to their world, and indications suggest that they were intelligent enough to have developed complex behaviors and social structures that are generally thought of only in the context of birds and mammals. Many dinosaurs, including the wiry *Velociraptor* and the long-legged ostrich-mimic dinosaurs, seem to have been well adapted for moving quickly. And even the diminutive *Sinosauropteryx* was fast enough to catch and consume lizards and mammals. Throughout their history, dinosaurs had been highly adaptable, and had maintained their role as the dominant land animals even though they were sharing their world with mammals, birds, crocodiles, turtles, lizards, and many other animals that still live with us today. Over the last few decades, the image of dinosaurs has been completely "made over," and it is no longer appropriate to consider them "stupid, unadaptable, and unprogressive."

Knight painted a pretty bleak picture of the end of the Cretaceous period, "with huge, bizarre and ungainly shapes rising and subsiding in the landscape, the earth covered with harsh and brittle scrub under tall palmettos, while both on the ground and in the trees, small, sinister, bright-eyed mammals awaited the slow but inevitable finale of the great Reptilian Era." In truth, most of the Late Cretaceous environments inhabited by dinosaurs were not dissimilar from many modern environments. Flowering plants had already appeared millions of years earlier, and the variety of habitats available for dinosaurs to live in was almost as varied as our modern world. Insects, fish, turtles, lizards, crocodiles, birds, and mammals would have been hard to distinguish from modern forms on the basis of superficial examination. Although dinosaurs would have appeared unusual to us at first, they in fact filled in niches that are today occupied in many parts of the world by equally bizarre animals. Think of the African veldt, for example, with a variety of species of crested duckbilled dinosaurs instead of antelopes and gazelles. Or herds of horned dinosaurs rather than the great herds of bison that used to cover the plains of North America. Is a giraffe any less bizarre than a long-necked sauropod? And was there anything so strange in dinosaurs as the elephant's nose? It was certainly a different world from any we know today. But it was not so foreign from modern environments that we could not relate to it.

During Knight's entire career, dinosaurs were considered to be reptiles that were best understood by studying modern reptiles such as crocodiles and lizards. To Knight and many of his contemporaries, dinosaur extinction made sense because "Nature had apparently grown weary of the great scaly cold-blooded monsters. They had been in existence too long. . . . And so the world was to grow away from these slow-moving dunces, and little warm-blooded beings, furry, alert and aggressive, were to supersede them."

Although the present may be the key to the past, it does not always unlock the door that you expect it to. It had long mystified people that dinosaurs, which were mere reptiles, could dominate the world for at least 140 million years. This was especially perplexing when one considers that mammalian history stretched back as far as or farther than the earliest dinosaurs. In other words, mammals were contemporaries of dinosaurs, and not just successors. With the beginning of the Dinosaur Renaissance (actually more like a "revolution"), paleontologists finally started to challenge the concept that dinosaurs were nothing more than overgrown lizards and crocodiles. We started to question whether or not dinosaurs really were cold-blooded (Bakker 1975; Desmond 1975). We became aware of more "intelligent" dinosaurs with brains as large as some mammals and birds (Russell 1997). We learned unexpected things about dinosaurian behavior (Horner 1997). The widespread acceptance of phylogenetic systematics, in conjunction with the fact that birds are probably the direct descendants of theropod dinosaurs, even leads us to question whether or not dinosaurs are extinct. Dinosaurs (if we include birds) are still more speciose than mammals, and in that sense are still more successful. Dinosaurs were different in many ways from modern lizards and crocodiles, and in some ways it is more appropriate to compare them with modern birds and mammals. But even that must be done with great caution, because dinosaurs were unique animals that were well adapted to the Mesozoic world, not to our world. Today, paleontologists tend to look at both birds and crocodiles, the two closest living relatives of dinosaurs, for clues to define the limits of dinosaurian biology. Dinosaurs may also, however, have had traits that were unique to them.

Life through the Ages is a wonderful retrospective view of science as it was more than half a century ago. Yes, there are the occasional errors, and it needs to be read with a grain of salt. But it helps us understand the mindset of an era when paleontology was still a young science. It is also a window into the mind of one of the greatest artists of prehistoric life. None can deny the tremendous impact his images had on the development of at least the public perception of the history of life.

People are probably more interested in Charles Knight now than they have been in the last half-century. He was even introduced as a fictional "character" in the Imax film entitled *T. rex: Back to the Cretaceous.* Knight's artwork, published here and widely available for the first time in five decades, should inspire a whole new crop of prehistoric artists!

References

Bakker, R. T. 1975. Dinosaur Renaissance. *Scientific American* 232 (4): 58–78.

Bakker, R. T. 1986. *The Dinosaur Heresies.* New York: William Morrow and Company.

Carpenter, Kenneth. 1999. *Eggs, Nests, and Baby Dinosaurs.* Bloomington: Indiana University Press.

Chiappe, L. 1998. Dinosaur embryos: unscrambling the past in Patagonia. *National Geographic* 194 (6): 34–41.

Coria, R. A., and L. Salgado. 1995. A new giant carnivorous dinosaur from the Cretaceous and Patagonia. *Nature* 377: 224–226.

Crumly, C. R. (ed.). 2000. *Dinosaur Imagery.* San Diego, Calif.: Academic Press.

Currie, P. J. 1998. *Caudipteryx* revealed. *National Geographic* 194 (1): 86–89.

Currie, P. J. 2000. Possible evidence of gregarious behavior in tyrannosaurids. *Gaia.*

Czerkas, S. Massey, and D. F. Glut. 1982. *Dinosaurs, Mammoths and Cavemen: The Art of Charles R. Knight.* New York: E. P. Dutton.

Czerkas, S. J., and E. C. Olson. 1987. *Dinosaurs Past and Present.* Vols. 1 and 2. Seattle: Los Angeles County Museum and University of Washington Press.

Desmond, A. J. 1975. *The Hot-Blooded Dinosaurs.* London: Blond and Briggs.

Dixon, D. 1993. *Dougal Dixon's Dinosaurs.* Honesdale, Pa.: Boyds Mills Press.

Erickson, G. M., and K. H. Olson. 1996. Bite marks attributable to *Tyrannosaurus rex:* preliminary description and implications. *Journal of Vertebrate Paleontology* 16: 175–178.

Farlow, J. O., and M. K. Brett-Surman. 1997. *The Complete Dinosaur.* Bloomington: Indiana University Press.

Galton, P. M. 1997. Stegosaurs. In J. O. Farlow and M. K. Brett-Surman (eds.), *The Complete Dinosaur,* pp. 291–306. Bloomington: Indiana University Press.

Heilmann, G. 1927. *The Origin of Birds.* New York: Dover

Books. Reprint, New York: D. Appleton and Company, 1972.

Horner, J. R. 1997. Behavior. In P. J. Currie and K. Padian (eds.), *Encyclopedia of Dinosaurs,* pp. 45–50. San Diego: Academic Press.

Huene, F. von. 1926. The carnivorous Saurischia in the Jura and Cretaceous formations principally in Europe. *Museo de La Plata, Revista* 29: 35–167.

Huxley, T. H. 1868. On the animals which are most nearly intermediate between birds and reptiles. *Annals and Magazine of Natural History* 2 (4th series): 66–75.

Patterson, B. 1959. *National Audubon Society, Nature Program, Prehistoric Life.* Garden City, N.Y.: Nelson Doubleday.

Paul, G. S. 1987. The science and art of restoring the life appearance of dinosaurs and their relatives: a rigorous how-to guide. In S. J. Czerkas and E. C. Olson (eds.), *Dinosaurs Past and Present,* vol. 2, pp. 4–49. Seattle: Los Angeles County Museum and University of Washington Press.

——. 1988. *Predatory Dinosaurs of the World.* New York: Simon and Schuster.

Russell, D. A. 1977. *A Vanished World: The Dinosaurs of Western Canada.* Ottawa, Ontario: National Museum of Natural Sciences.

——. 1997. Intelligence. In P. J. Currie and K. Padian (eds.), *Encyclopedia of Dinosaurs,* pp. 370–372. San Diego: Academic Press.

Tanke, D. H., and P. J. Currie. 2000. Head-biting behavior in theropod dinosaurs: paleopathological evidence. *Gaia.*

T. rex: Back to the Cretaceous. Imax film.

Life through the Ages

Foreword

During the closing years of the eighteenth century Baron George Cuvier, a noted French scientist, advanced some extraordinary theories on the subject of life upon this earth and how it had been developed through a very long period of time. His main idea, embodied in the word *catastrophism,* imagined a series of recurrent and terrible upheavals of the earth's surface during which all the life of a certain period was utterly destroyed. He also believed that when conditions again became favorable, a wholly new and more advanced set of animated beings would arrive to take the place of the former unfortunates. Thus, life would go on from age to age, ever changing, ever becoming more perfect. Surely, this point of view was a fantastic conception of what really occurred, but so great was Cuvier's influence upon the scientific thought of his time that for years, apparently, no one dared to put forth a different notion in regard to this profoundly important question.

At length, in the person of Jean Baptiste Lamarck, a champion appeared who did not fear the mighty Cuvier nor his transcendent prestige and who was bold enough to declare his personal opinions on the deeper meaning of life and the steps whereby he had formulated his conclusions. In direct contrast to Cuvier's ideas, Lamarck believed that life had been continuous since its first inception, but that new forms were constantly being altered and shaped by their surroundings. One may imagine that this idea was regarded at the time as being completely revolutionary, but in spite of much opposition it became more and more a favorite theme in the minds of scientists deeply interested in the subject. When, some fifty years later, Charles Darwin, the great exponent of the theory of evolution, published his epoch-making volumes, *Origin of Species,* and *The Descent of Man,* it was realized that Lamarck's thesis had fallen upon good ground and that, despite many additions and alterations, Darwin had adopted the principles of Lamarck's wonderful presentations. In the main, modern thought subscribes to similar conceptions of life in the world about us.

In this book I have shown examples not only of the more striking forms of Prehistoric times, but in many instances, their modern representatives, in an effort to bridge the gap between the past and the present. Special attention will be devoted to this phase of the subject in the captions opposite each picture.

Cambrian Epoch

The name Cambrian was derived from the Roman word Cambria applied to the region in Great Britain, now known as Wales.

The picture is new, but the underwater scene which it represents is perhaps some 500,000,000 years old. Nevertheless, looking at it closely, one sees and recognizes certain forms of life which differ but slightly from those of our own time. Jelly fishes are floating lazily through the clear warm water, shrimp-like creatures dart about in search of food or to escape their enemies, and graceful seaweeds and sea lilies wave their delicate fronds to and fro in response to the passing current. These so-called sea lilies are, of course, not lilies at all, but a kind of upturned starfish with a stem projecting from the under side by means of which they fasten themselves to the sea bottom. There are sea worms too, in the picture, not unlike those in existence today. The largest object, the crustacean Sidneyia, prone upon the sand, has vanished completely, as have also the queer floating basket-like variety of shellfish, and the two oval creatures in the middle distance known as Trilobites. These belong to a race of marine animals which flourished in vast numbers for ages without very much change of form.

There is still a good deal of dispute as to their actual affinities, some men claiming their relationship to the crustaceans, others to the king or horseshoe crab of our summer beaches. You have doubtless come across this latter creature, well-named for its peculiar shape, as it crawls slowly over the sandy bottom, close inshore, in search of food. If we raise this grotesque object by means of his sharp spine-like tail, we will be surprised at the wriggling legs on his underside. Strange visitor from the past, he takes us back to the far reaches of those distant days. Curiously enough, the wonderful little animals seen in the picture, and many more besides, were discovered beautifully preserved, some 10,000 feet up in the Burgess Pass of the Canadian Rockies. Dr. Charles Wolcott, later Secretary of the Smithsonian Institution in Washington, D. C., came across them on one of his many trips in search of fossils. Once on the sea bottom, time and earth convulsions raised this interesting little host to their present great elevation above sea level. Their discovery opened up a whole new world of ancient creatures, whose story added one more chapter to the Book of Life.

Though we begin our book with the Cambrian Period, life itself must have been going on for vast eons before that day, because all the species shown were well-perfected and adapted to their particular part in the scheme of things.

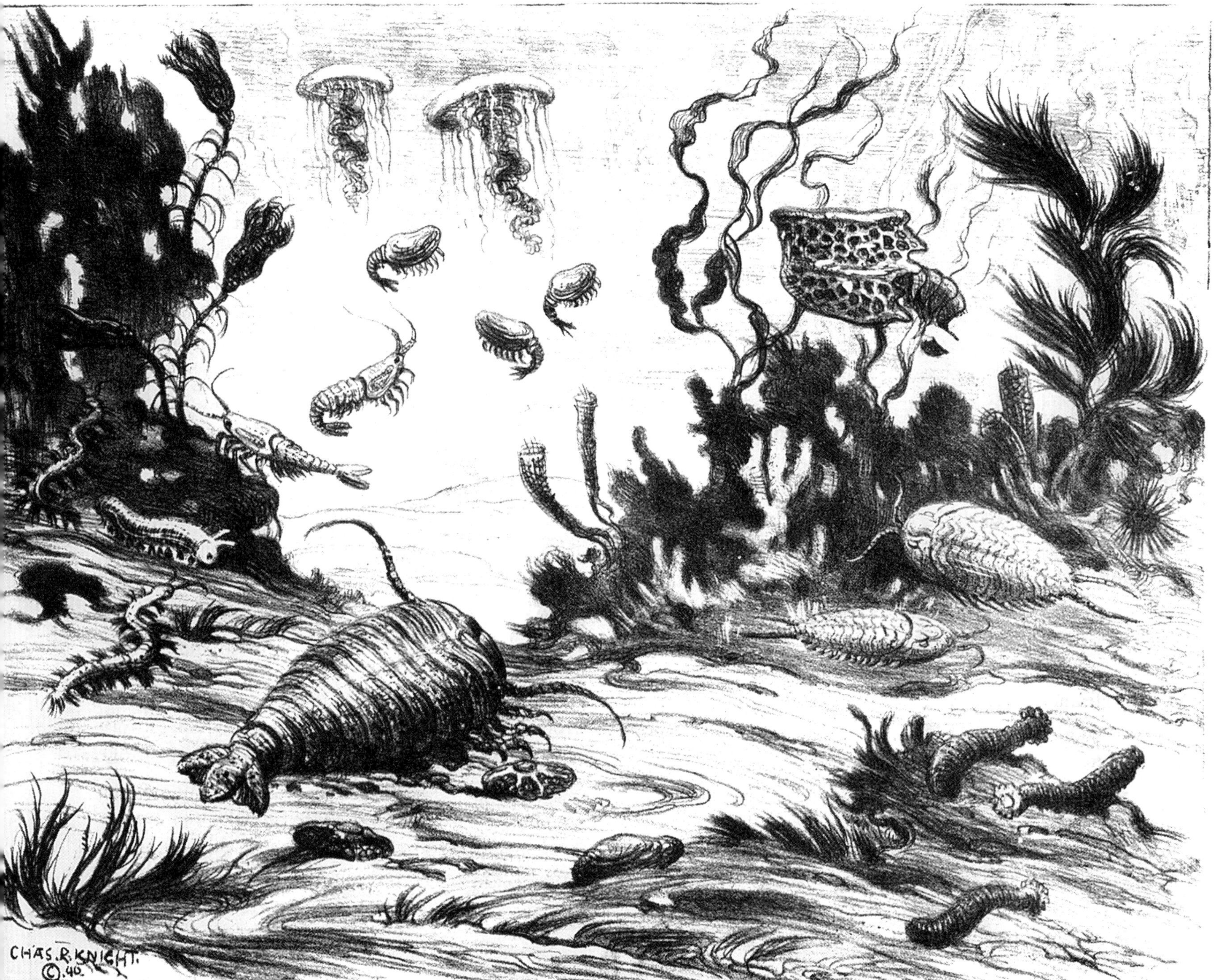
CHAS. R. KNIGHT.
©. 40.

Devonian Period

The name Devonian is derived from the county of Devon in England where many of the fossils were discovered.

No spiral-shelled creatures with long waving arms, no finned and scaly shapes floated through our Cambrian ocean, but as ages passed life in the great deep continued to change. We now see other and more complicated forms, apparently all a part of Nature's great plan of evolution. How or why they were so developed we cannot truly say. Enough for us that through the clear blue depths, glittering graceful bodies move with assured ease, a race newly risen from the ocean floor on which can be seen several of the still grovelling, awkward forms of their more primitive relatives. At first glance, these lowly little armored things seem hardly fish-like, but on looking more closely we can make out distinct affinities with their more aristocratic descendants. However, even these will gradually change and will eventually produce the gleaming beauties of our modern seas—tunas, swordfish, the brilliant reef species and hosts of other fairy-like creatures filled with life and energy. And what of the sinister brute with the coiled shell and the bright hard eyes which lies so quietly under the waving sea lilies? Around its head are many twisting feet or arms and so it is called a Cephalopod (head-footed). It is a mollusc, and one of a long line of extraordinary animals, all fierce, carnivorous and repellent. In various eras the coiled shells straightened temporarily, but the beautifully curved form again is present in the recent pearly nautilus, a deepwater resident of the Indian Ocean. In contrast to these protected species, the great-eyed modern squids, which dash madly through the open water after their finny prey, have no outer shell. The ten suckered tentacles projecting from the head—two of which are much longer than the others—seize and hold the unfortunate quarry, drawing it back to the parrot-like beak in the center of the ring of arms. Even more terrible in aspect is the octopus. As befits its eight-armed form, it crawls or swims slowly about seeking its prey of crabs and lobsters, the flesh of which it devours, discarding the hard outer shell. By comparison, the pretty little female Argonauts sail gaily along the surface of the sea, their bodies partly concealed in a glistening boat-shaped shell. These, as well as the cuttle fishes, throw out a blackish substance, called sepia. A pigment of rich brown color prepared from this secretion was used by artists especially in water-color painting. These head-footed octopuses are molluscs though it is difficult to see why they should be classed with clams, oysters, scallops, and sea and land snails. Nevertheless, technically, they are so connected, and as a great division of the animal kingdom they have an incredibly long life history.

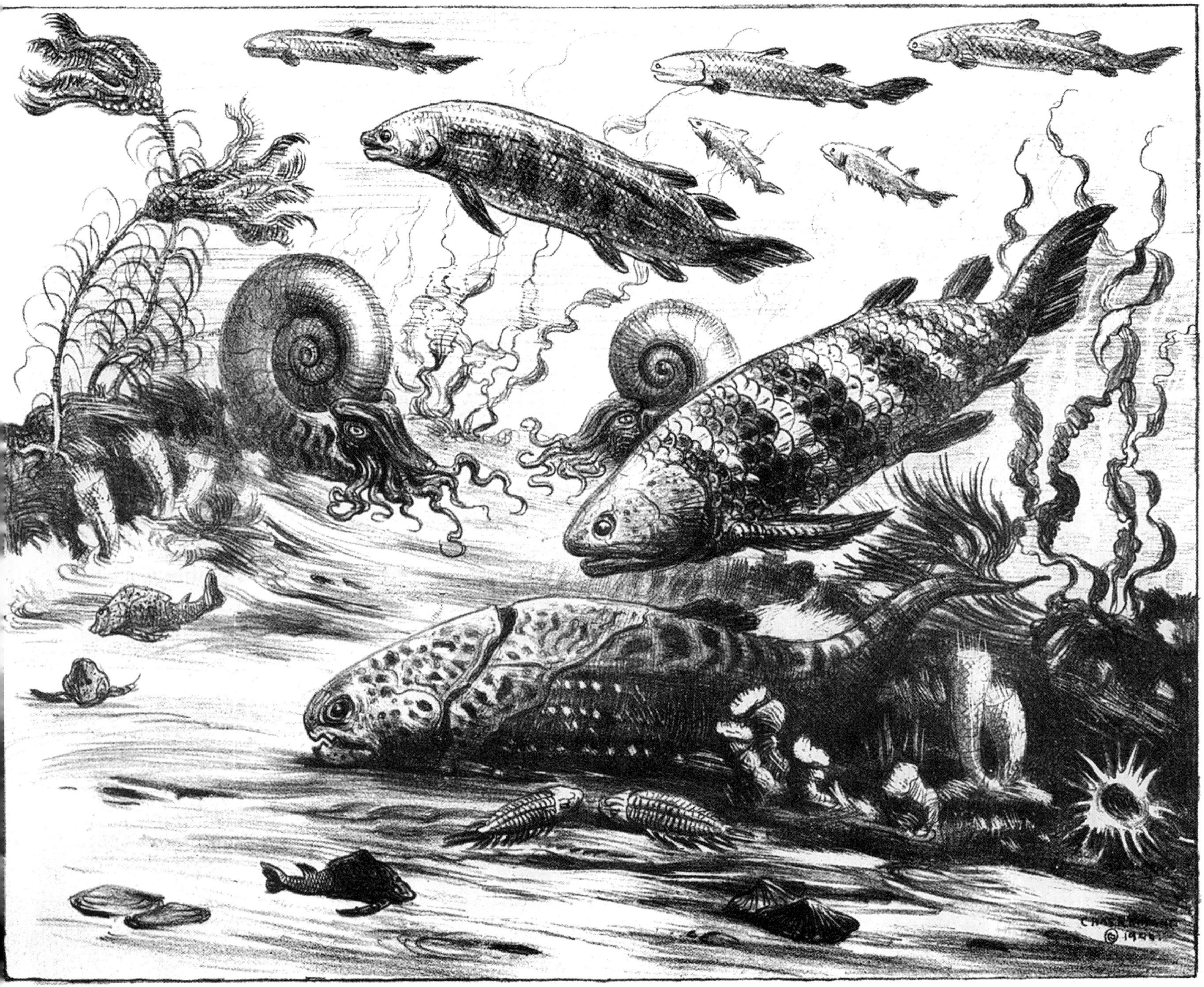

Carboniferous or Coal Period

A gentle breeze blows softly through the forest glades; the silence is oppressive, for no song of birds, no cry of an animal breaks the stillness of that shadowed land. Strange trees lift their stumpy branches to the dark and lowering sky; silver threads of water gleam on every side, while fearsome pools reflect the foliage above them. Suddenly, a line of ripples falling back from a broad flat head cuts the quiet surface of a stream. Huge dragonflies soar gracefully in the open spaces and cockroaches, big as mice, scuttle busily about over the leaves and moss-covered logs of the forest floor. For in this Carboniferous Age, the chief coal-making period of the world, life had so far advanced that the more progressive species were able to emerge from their watery home and crawl about upon the land. They breathed the air through lungs, not gills, and used their little feet and legs instead of fins or paddles as a means of progression. Great sprawling salamander-like creatures such as Eryops are typical of the period. No doubt these and many other species still passed their earlier life stages in the water. They were stupid, smooth-skinned monsters, some six feet long, with wide tooth-filled jaws and an enormous gape which enabled them to swallow their food at a single gulp. They probably avoided the sun as it was too hot for their delicate moist skins, so that the prevailing dark and gloomy climate of the time must have suited them perfectly. It was an era of swamps, thick foliage, and huge rushes, thirty feet high, which rose in grotesque clumps from the slime-covered morasses. As decay and death overtook these forest giants they eventually crashed into the mud and ooze surrounding them. There, time and pressure forced them to sink ever deeper into the muck and mire, where, after thousands of years, their trunks formed the substance we now call coal. A profound change in the life of both animal and vegetable kingdom was to occur after this long continued humid period. The climate became drier, and a great new family, the *Reptilia,* made their appearance.

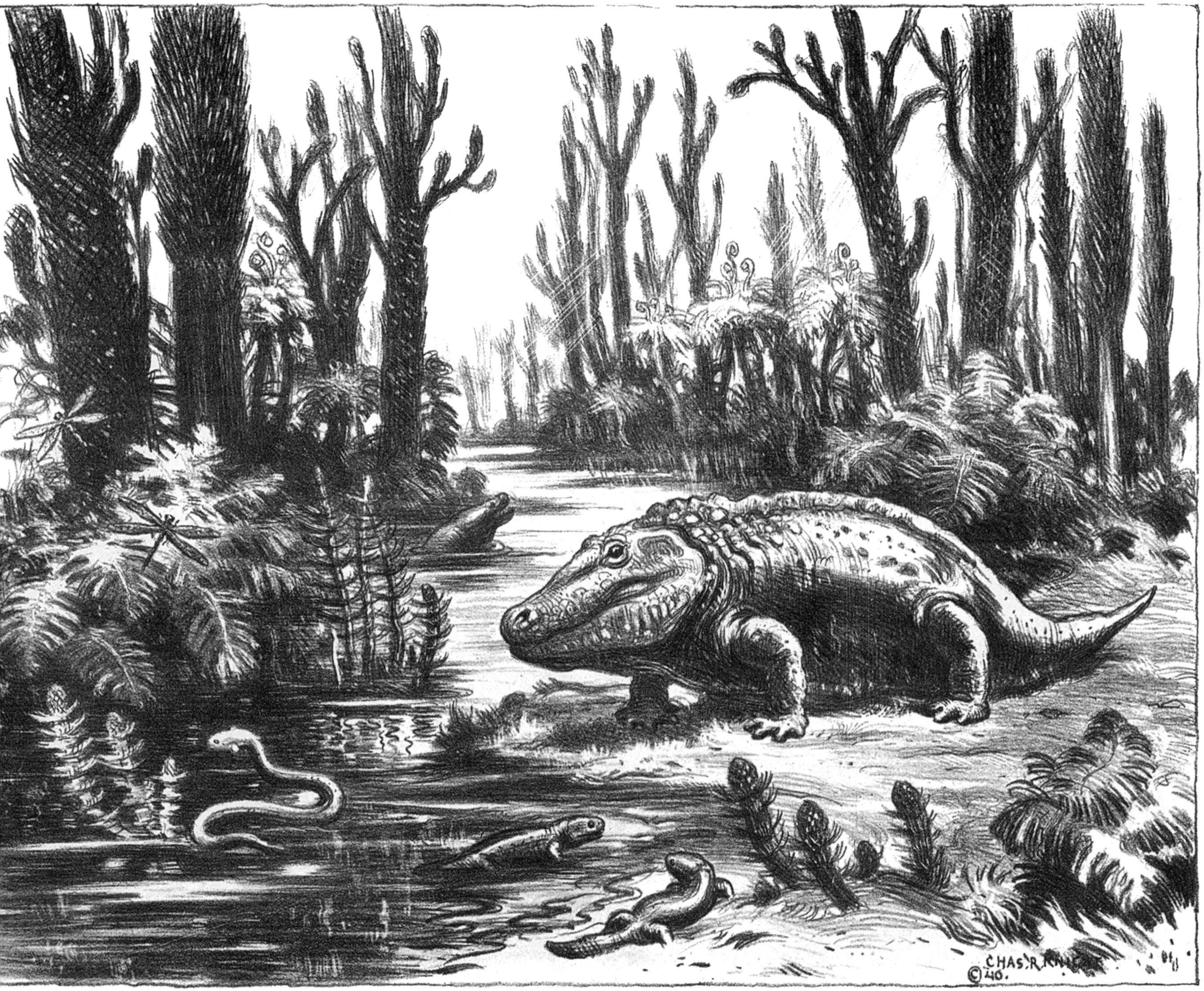
CHAS. R. KNIGHT
© '40.

Dimetrodon—Permian Time

(NAMED FROM THE PROVINCE OF PERM IN RUSSIA)

How utterly unlike our previous picture is the one before us now! How stiff and firm and regal is the unique reptile with the huge, finlike crest along its back! Here is no cringing, tender-skinned salamander, lurking in the dense shade of some old tree, but a true reptilian form, taut, and scaly, a ferocious killer if we may judge from its long sharp fangs. Some six feet of this extraordinary creature now presents itself to our attention and it would seem that no queerer reptile ever crawled upon this earth since the beginning of time. Naturally, we would like to know the use of the immense, bizarre back ornament, but there is no answer to that query. This strange development is not a fin but an elongation—beyond all natural limits—of the vertebral spines. It stands up rigidly some four feet in the middle of the back and shortens quickly at either end. A tough skin must once have joined all the delicate points, but even so, the arrangement would seem to have made the creature very vulnerable to attack. Dimetrodon was only one of scores of odd-looking reptiles which flourished in many widely-separated parts of the world during Permian time. The climate in contrast to that of the coal period was dry. There was more sunlight and less luxuriant vegetation—a fortuitous combination for the growth of tremendous reptile types. It made necessary a special skin formation unlike the moist delicate covering of the salamanders and vastly more resistant to intense sunlight. The scaly, hard-skinned reptiles prospered under these conditions, so that this kind of cold-blooded creature grew apace. They gradually separated into a multitude of different forms which were able to dominate the world for ages. These included huge dinosaurs, flying pterodactyls, and lithe, powerful marine species, icthyosaurs, plesiosaurs, mosasaurs, as well as crocodiles and turtles. Strange indeed, must have been a landscape peopled by these grotesque beings—bright-eyed and ever hungry, without fur or feathers to soften their harsh and repellent contours. We need have no qualms about meeting this scaly crew for most of them have gone forever, some 50,000,000 years ago, leaving only crocodiles, turtles, lizards and a few snakes to remind us of the grand old Age of Reptiles.

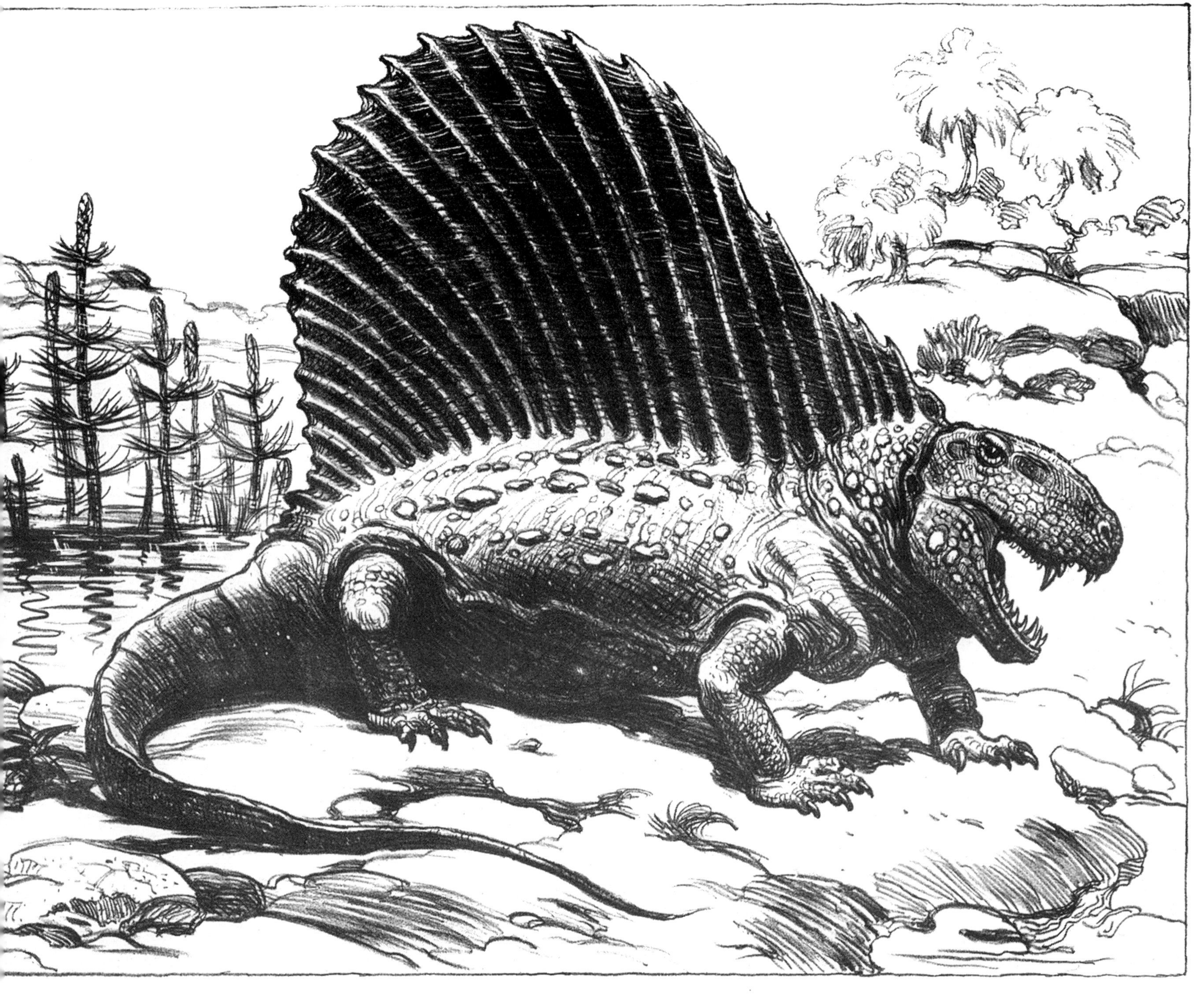

Reptiles from the Karoo Region of Africa

Later, in our series, we shall see pictures of dinosaurs and the birdlike forms which we think may have descended from them. In the picture some possible connection will be shown between certain solid-skulled, queer-looking reptiles and true warm-blooded mammals. Fossils of these most interesting variations of the reptile type are now found in at least three widely-separated areas of the world—Texas, Siberia, and the Karoo region of South Africa. The species in the two latter places are quite similar. There are both flesh-eating and vegetarian varieties. Some of them quite closely resemble our primitive modern mammals, especially as to the form of the teeth. It may interest us to know what we consider a primitive animal. One of the best-known is the opossum, famous in the Negro folklore of our country and a sworn enemy of roosting chickens. To us he may seem just another animal, but scientific enthusiasts consider him a kind of leftover from the past. Because many of the features in his anatomy are similar to those of primitive animals, the comparison to an old Permian reptile, Ictidopsis (weasel-faced), is striking. Of course, the opossum has hair and the reptile, scales, but even so, the big heads with their long, delicate jaws and sharp teeth, are very similar. Such a comparison is of great interest to those who seek connecting links between one great division of animal life and another, for they explain much that is mysterious in the history of created beings—their gradual evolution from lower to higher development, and their increase in brain power so necessary in the maintenance of life in any given set of circumstances. The huge-beaked reptile at the top of the drawing, which at first glance looks like a big lizard, is still another kind of mammal-like creature, a vegetable feeder. Scientists think that when hair gradually replaced the scales on the body, the temperature of the blood became higher and this stimulated quicker action, keener mentality, and promoted greater adjustment to climatic conditions. Slowly but effectively, the newly-formed mammals which were a progressive type of the more ancient reptiles, were able to take their place in the world and to multiply. They increased in size and variety and spread to all parts of the earth. So, in our present-day opossum are found suggestions of the structure of these creatures of an earlier age.

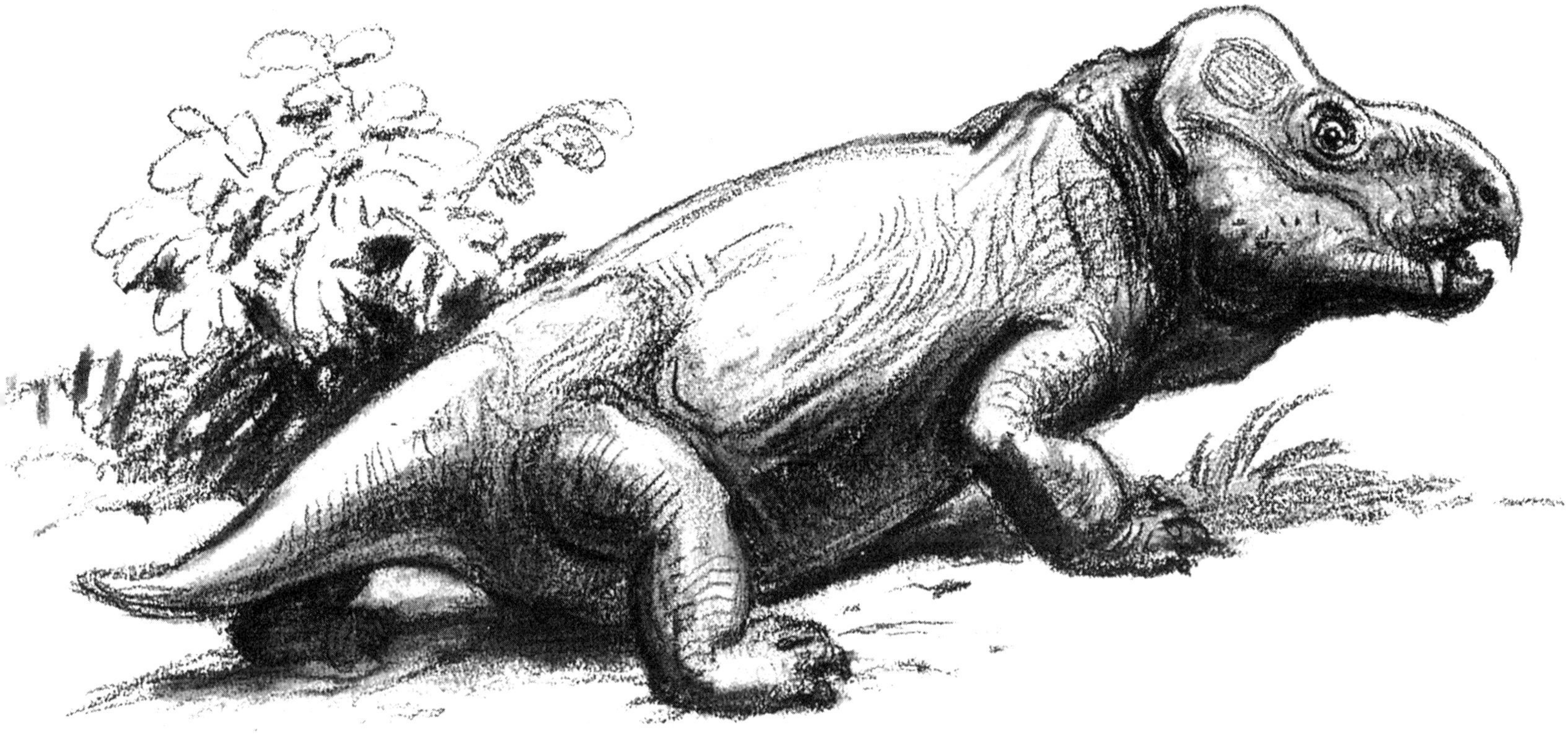

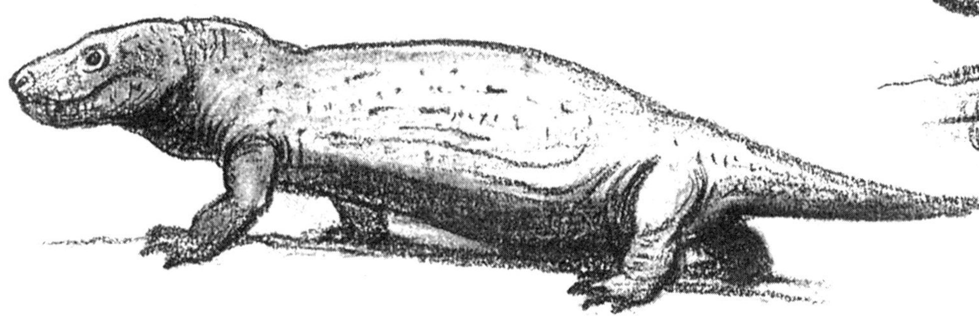

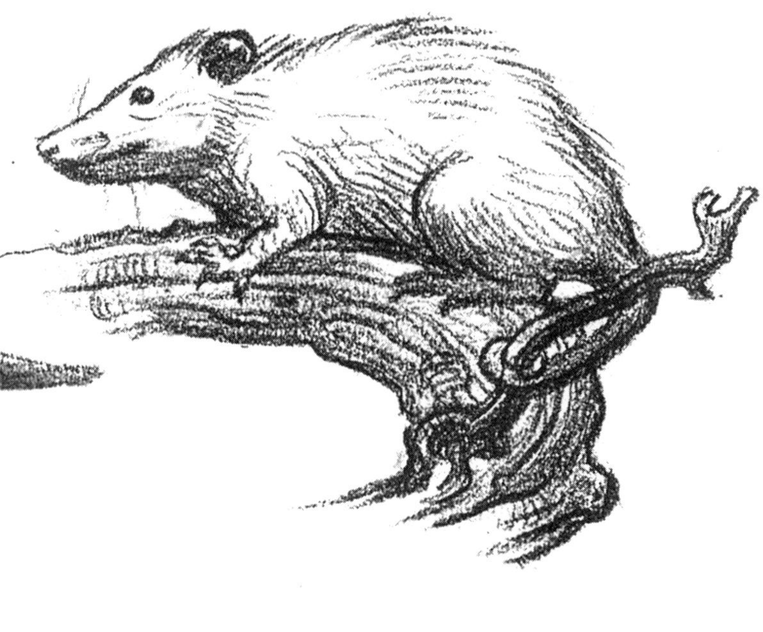

Allosaurus—Jurassic Period

(NAMED FROM THE JURA MOUNTAINS)

The Allosaurus, with big, curiously birdlike hind legs and feet, and small front limbs armed with long, sharp claws, is a true land-living type of dinosaur. The head was large, the jaws long and filled with sharp-cutting teeth. He used his long, heavy tail as a balance while walking. He was one of an old line of similarly-formed reptiles, some of which were quite small, while others, in later eras, were almost fifty feet in length. In the picture, the fierce brute is crushing to death a small individual of the Brontosaurus family, a water-living form, which he may have seized as it swam slowly along the shores of a sandy lagoon. While he preferred living food, he would no doubt eat carrion, as is proved by tooth marks on the vertebrae of a large Brontosaurus, too big for even Allosaurus to tackle in a fair fight. Twenty-five feet long, lightly constructed for quick action, and fairly sagacious for a reptile, this savage killer secured its prey very easily, pouncing upon some hapless victim and biting and clawing it without mercy until it died. At first glance, one might suppose that these long-legged, birdlike dinosaurs hopped about like kangaroos, but this idea of their mode of progression is altogether wrong. The rather stiff and ungainly creatures really walked, or ran as do most birds, moving forward first one leg, then the other. A plucked turkey with a long tail and little front legs with claws instead of wings, conveys a pretty good idea of what all these flesh-eating reptiles looked like. Indeed, as we study them closely, their birdlike affinities become more and more apparent as we shall later see in the Archaeopteryx (the bird reptile), a sort of little dinosaur with wings instead of front legs, and feathers, which further enhanced the birdlike appearance. Allosaurus flourished in a climate not unlike that of Florida today—fairly dry, almost desert-like at certain seasons.

This period, the Jurassic, is known also as the Age of Cycads, a species of which is now known as the sago-palm. Vegetation generally was sparse. Later, however, the flora grew more luxuriant and true palms, bananas, sequoias, and gingko trees lent a more gracious aspect to the landscape. And in that landscape, larger, queerer, and more terrible creatures than any I have shown you, walked and ate and fought and died.

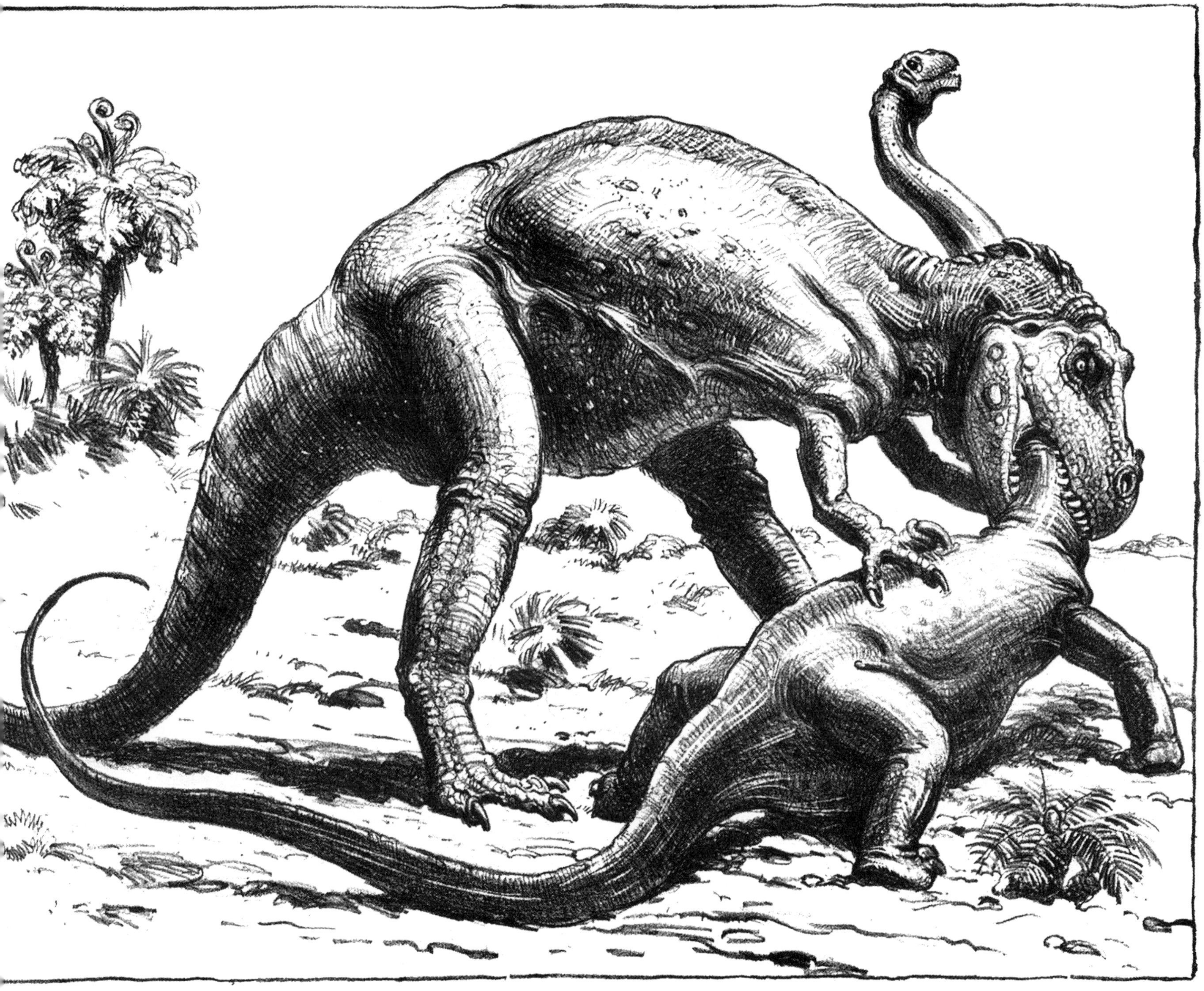

Brontosaurus—Jurassic Period

If some of the creatures already described have been extraordinary in physical appearance, what shall we say of the enormous brutes whose long necks and tiny heads rise so majestically above the water in this Jurassic scene? Some seventy feet of this big-bulked and small-brained Saurian unfolds itself before our astonished gaze. Brontosaurus, (thunder lizard), was a magnificent example of the water-living dinosaurs. It spent practically all of its time in the water, feeding on soft vegetation which it bit off with its small finger-like teeth. It could wade or walk on the bottom in twenty feet of water, its long neck and head moving slowly about in search of food, while the small eyes kept watch for possible land-living enemies, in the shape of some carnivorous species, patrolling the shores and beaches. By this time, the great dinosaur tribe had been on earth for ages, and was now separated into various species and families. Some were adapted to life in the water, others were land-living flesh eaters, and still others, waders. Brontosaurus, on account of its size, lived almost continually in the water and probably had living young; whereas Stegosaurus (roofed lizard), the strange being on the spit of land at the left of the picture, was distinctly a land animal. A double series of high, thin plates rise along the back of this fantastic twenty-five-foot monster, and four-pointed spines grace his long flexible tail. The minute head seems merely a nub at the other end of the body. Stegosaurus was, no doubt, the stupidest member of a very moronic family. He was, however, a fine example of Nature's method of so arming her really harmless children that merely to look at them was enough for the average flesh eater, who immediately lost his appetite and looked elsewhere in his search for food. Down through the ages creatures of every sort and kind, both in the sea and on the land, have assumed various terrifying outgrowths of scales, horns, etc., by means of which they seem to have been able to escape annihilation. As a rule, the true killers are armed with nothing more than teeth and claws, and with these weapons they battle their way through life. Their only enemies are old age and the consequent inability to secure food. Starvation is their greatest bugaboo.

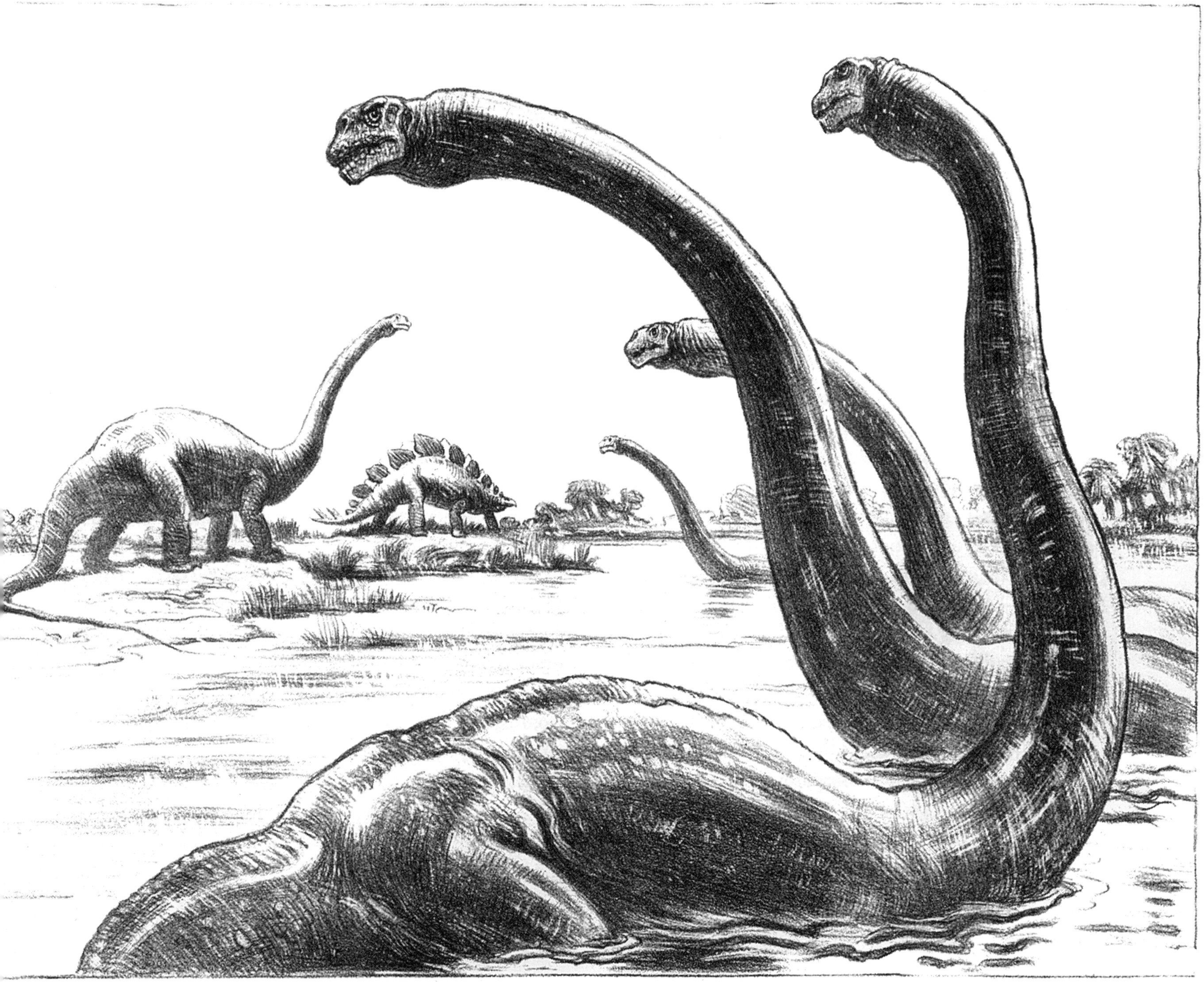

Archaeopteryx and Pterodactyl—Jurassic Period

In the long history of the past, Nature has devised many forms of locomotion for her precocious children—swimming in the water, walking on the land, and most difficult of all, the art of flying through the air supported on wings. In Archaeopteryx, the curious birdlike creature discovered some years ago in Bavaria, we see embodied one of these first attempts at flight. It was not wholly successful, but it was the forerunner of a long line that eventually led to the graceful, powerful birds whose wonderful aerial accomplishments we never tire of watching. If ever a treasure was unearthed among all the various remains that have come down to us through the years, surely it is Archaeopteryx, the answer to the fossil-hunters' prayer. For this little creature, no larger than a pigeon, is a link between the long-tailed scaly reptiles on the one hand and the feathered birds on the other. A pair of feathers depended from each tail vertebra and it had claws on its wings. Though it had no true beak, its jaws were lined with small sharp teeth. The feet resembled those of birds. Sparse feathers partially covered the body, but the weak wings could do no more than sustain the delicate little being as it fluttered about among the branches or planed lightly to the ground.

Just beneath the Archaeopteryx is a weird and oddly-constructed creature, a Pterodactyl (wing fingered), which had another kind of flying mechanism. It was a very efficient one, for it enabled this absurd little reptile to get about at good speed, catching insects and other prey while on the wing. Its wing was peculiar and very different from that of a bird or bat, as the long drawn-out point was the lengthened little finger, the other fingers remaining as claws. From the tip of this digit a membrane extended back to the knee of the hind leg, forming a long soaring surface for the animal in its travels. In size these leathery-skinned flying reptiles varied from tiny ones no bigger than a robin, to great airplane-like creatures twenty feet across the wings. Long ago the Pterodactyls vanished from the skies but the descendants of the little bird-reptile now compete with the bats and the insects for aerial supremacy.

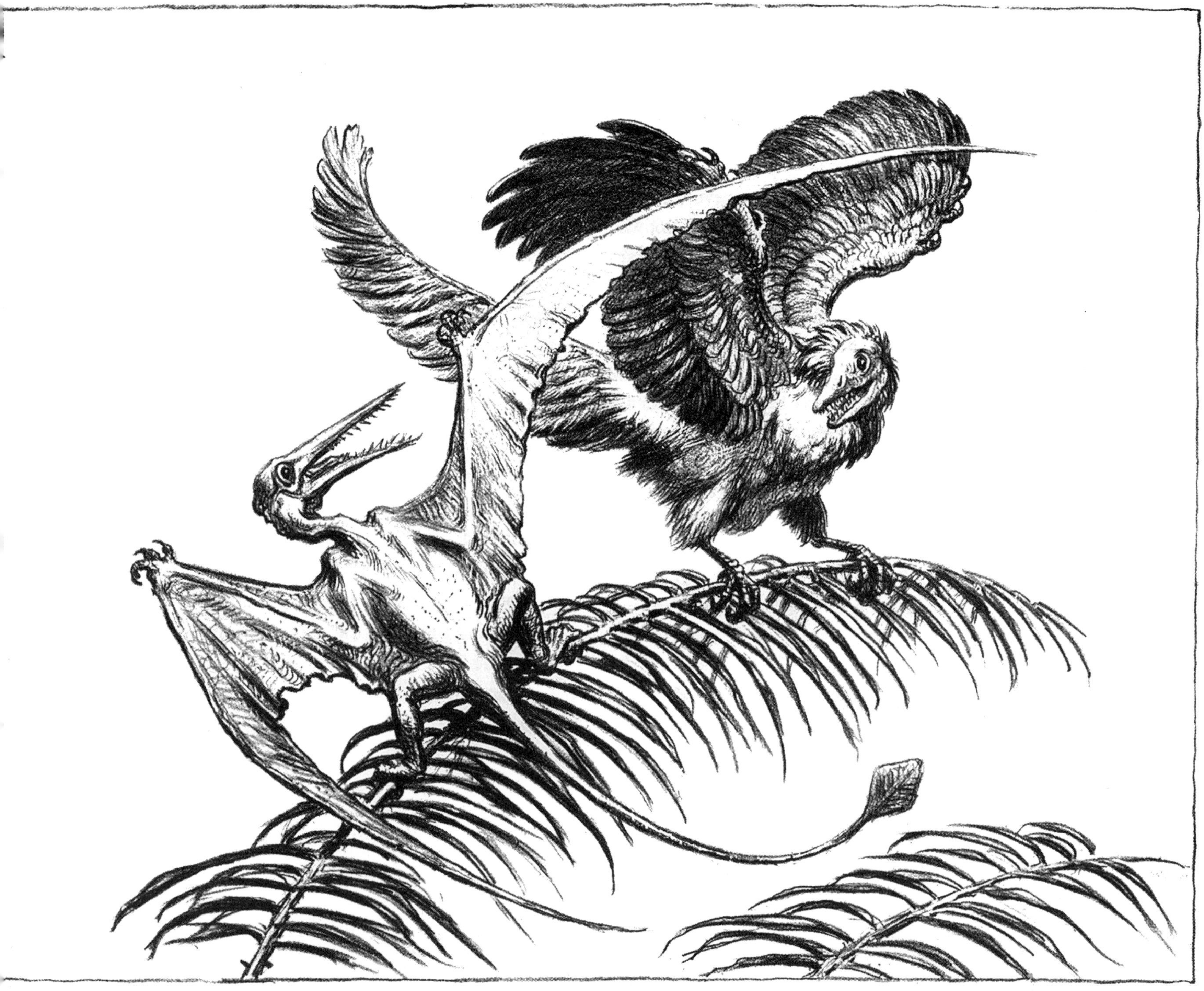

Mosasaurus—Cretaceous Era

The surface of this earth is not and never has been perfectly stable, with the land and water areas constant in size and relation to each other. On the contrary, through long geological periods tremendous changes have occurred. So great were these that at one time a warm shallow sea cut through the western part of North America from the Gulf of Mexico to the Arctic Ocean. This body of water fairly teemed with fish and great marine reptiles. Among these, Mosasaurus, (Meuse Lizard), was prominent. It was a huge, long-jawed creature whose thirty feet of sinuous body and tail, equipped with four great swimming paddles, would have given rise to all sorts of sea serpent stories were it in existence today. Sea-bathing on the Kansas beaches would have been a dangerous game, as this massive marine lizard could have swallowed a man at a single gulp. Other species of big reptiles also swam these tepid seas and as they died their limp bodies sank into the fine silt at the sea bottom, eventually becoming petrified skeletons, with every bone in its right place. Today, after the surrounding matrix has been removed, we gaze in amazement at their huge forms and picture to ourselves the living animals as they dashed through the cretaceous seas fifty million years ago. As none of these fossil types could come out of the water to lay their eggs, they probably all had living young, born (as are many of the sharks) perfectly formed and able to shift for themselves.

Above the sea monster, soars a great Pterodactyl (Pteranodon), with a twenty-foot wing spread. This species differs from the one previously described in that it had sharp pointed jaws with no teeth, and a long projection from the back of the head—a rudder, no doubt—in lieu of the usual long tail. How this weird being rose from the water if he once settled upon it, is difficult to understand, for his breast muscles were weak and he was unable to flap his wings. With the passing of this great species the race of flying reptiles came to a close.

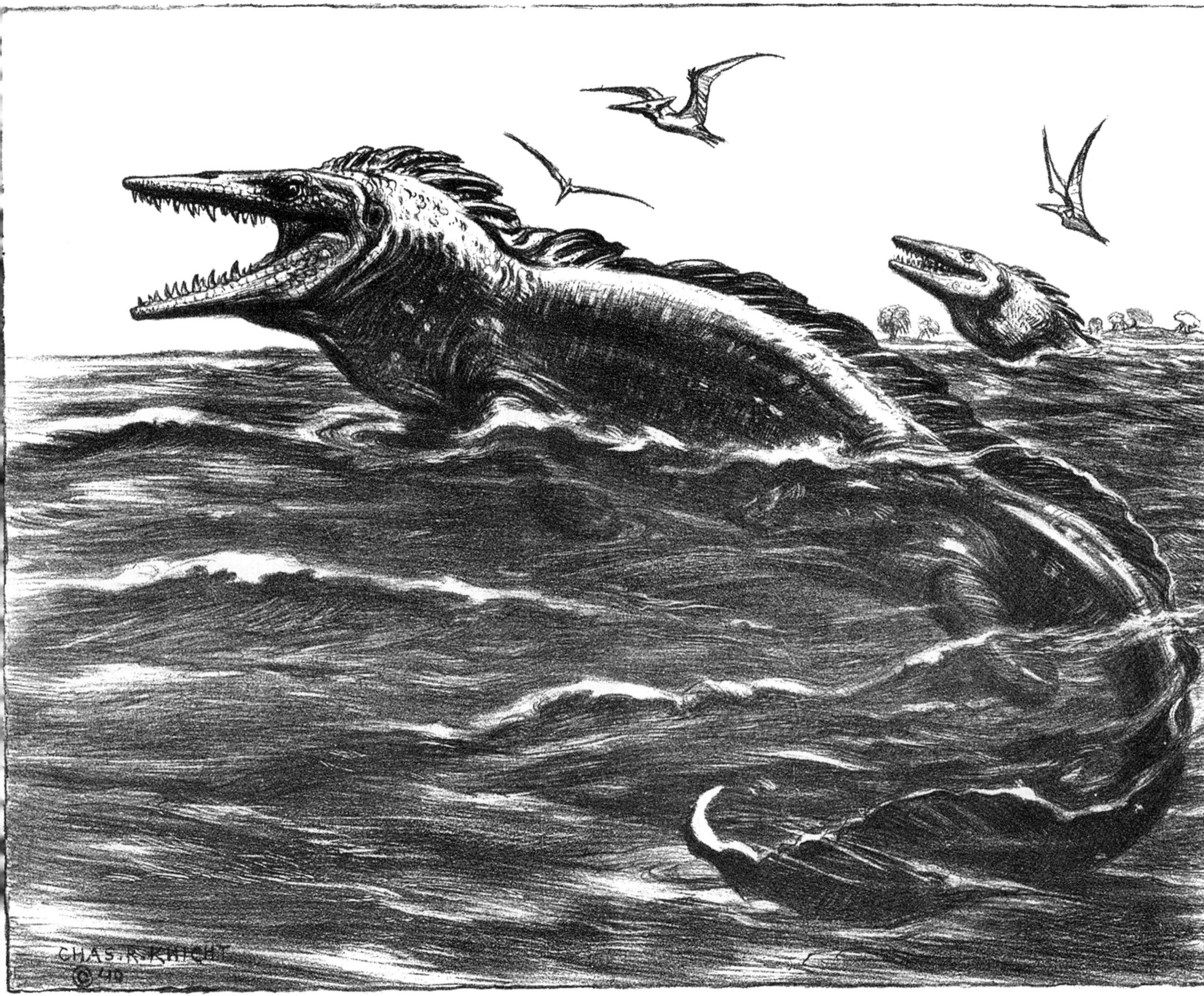
CHAS. R. KNIGHT
© '40

Styracosaurus—Cretaceous Era

"I'd hate to meet him in the dark." It's an overworked phrase but it applies to this very extraordinary-looking dinosaur better than to most things. However, appearances are deceptive, and Styracosaurus was really a harmless old fellow, a vegetable feeder, armed only for defense against marauding flesh eaters. That enormous collar of spikes sticking out around his head, together with his nose horn and sharp turtlelike beak, no doubt turned away many a hungry Tyrannosaurus or other brute, which otherwise would have given him a few terrific bites on the back of the neck and so finished him in short order. A big brother in ugliness, Triceratops, had a small nose horn and two very long pointed horns, one over each eye, as well as an enormous bony collar without spikes. Both these titanic creatures came late in Dinosaur time, at the close of the Cretaceous Era. They had to be protected against huge flesh eaters. All species were by now too big and too ungainly for their own good and though they didn't realize it, were doomed to pass away completely, for Nature had apparently grown weary of the great scaly cold-blooded monsters. They had been in existence too long, for they were stupid, unadaptable, and unprogressive. And so the world was to grow away from these slow-moving dunces, and little warm-blooded beings, furry, alert and aggressive, were to supersede them. As a matter of fact, these little creatures had been right under their ponderous feet for ages and had even destroyed their eggs on occasion. The great Dinosaur family, however, was hard to eliminate completely, and for a time, at least, various species may have roamed the great open spaces in many lands around the globe. Heavily armored, tank-like varieties ambled about, and long-legged, semi-aquatic types with heads like ducks, fed along the edges of the waterways. These dreadful-looking but inoffensive beasts did their best to stave off the ever voracious meat eaters who stalked about in constant search for food. It must have been a most depressing world as we think about it now, with huge, bizarre and ungainly shapes rising and subsiding in the landscape, the earth covered with harsh and brittle scrub under tall palmettos, while both on the ground and in the trees, small, sinister, bright-eyed mammals awaited the slow but inevitable finale of the great Reptilian Era.

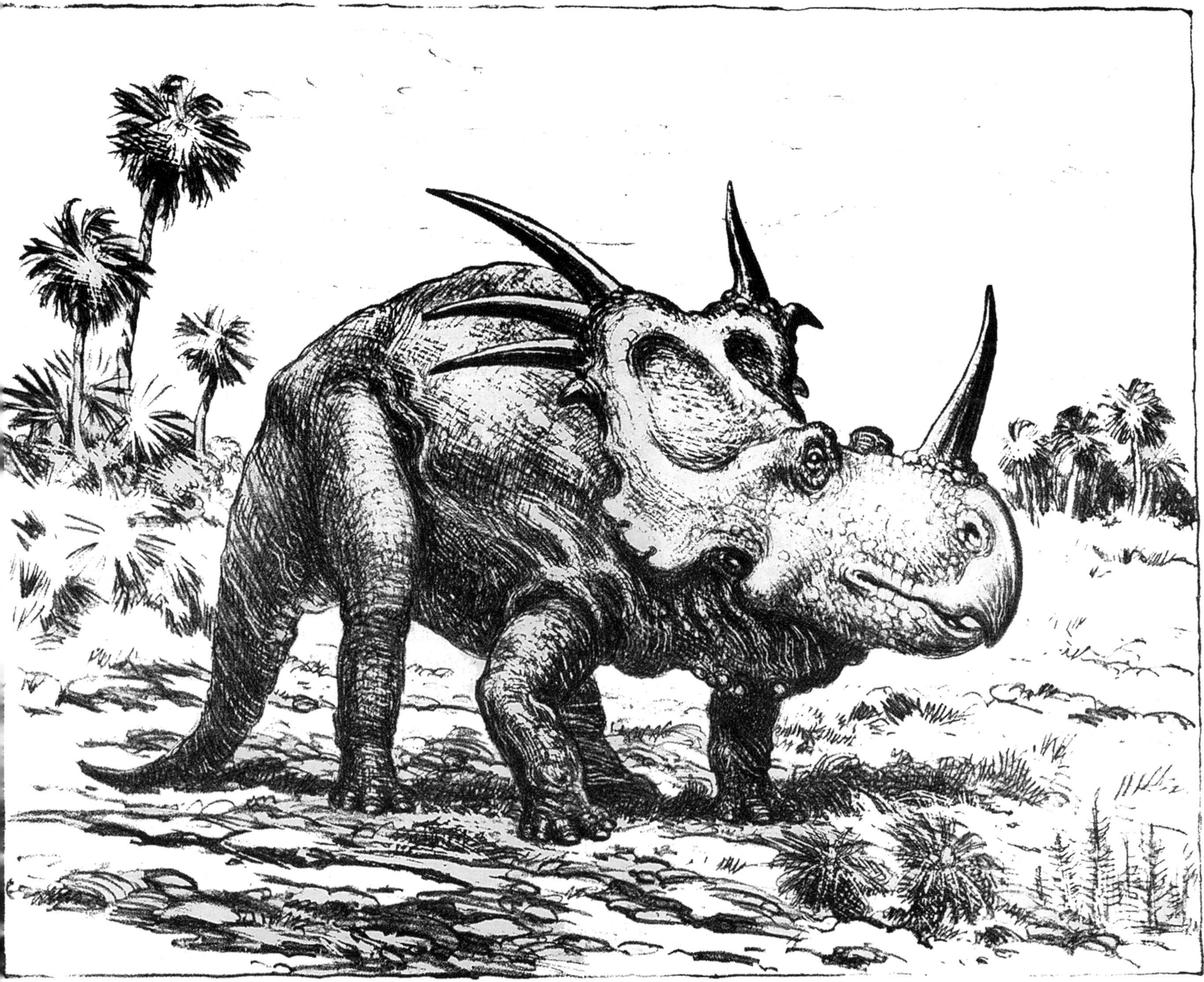

Tyrannosaurus, King Tyrant Lizard—Cretaceous Era

We have saved the grandest and fiercest dinosaur for the end of their life-story because, as the last tremendous example of a mighty race, he really belongs there. He towers above us, the terrible head, perched on the short, thick neck, rising twenty feet into the air. As we follow his contour down to the vast hind legs and birdlike feet, to the end of the twenty-foot tail, horror grips us, so terrifying is our first impression of this mighty creature. A glance at the powerful jaws lined with rows of sharp-cutting teeth, and at the tiny front legs and claws, scarcely visible against the body bulk, distinctly adds to that first vivid impression. From his great height the bright far-seeing eyes could encompass a wide stretch of country, and woe betide any unfortunate vegetable-eating dinosaur within their scope, unless it happened to be a large and powerfully armed *TRICERATOPS,* (he of the three-horned face). In that case the Tyrant, faced by the huge but really harmless horned monster, might think twice before attacking such a formidable-looking object. One might be excused for being a bit facetious about this great brute simply because he is so dreadful in appearance. He was just an enormous eating machine with an insatiable appetite and with practically no brains. When he really did tackle some helpless species of dinosaur, a few vicious bites would be inflicted across the back, and the victim, quickly torn to pieces, was promptly swallowed. The King Tyrant himself was the culmination of millions of years of dinosaur evolution. From the Triassic through the Jurassic, Allosaurus stage, to his present massive size, he and his close relatives—the flesh eaters, big and little—had all been fashioned along the same lines, a make-up that was evidently a successful one. Meanwhile, in an attempt to circumvent just such killers as Tyrannosaurus, the vegetable-feeding species had branched out into all manner of grotesque and awe-inspiring shapes. In any case, the whole scaly, spiny lot have long since vanished, which perhaps is just as well, because no more sinister beings ever walked the surface of this earth.

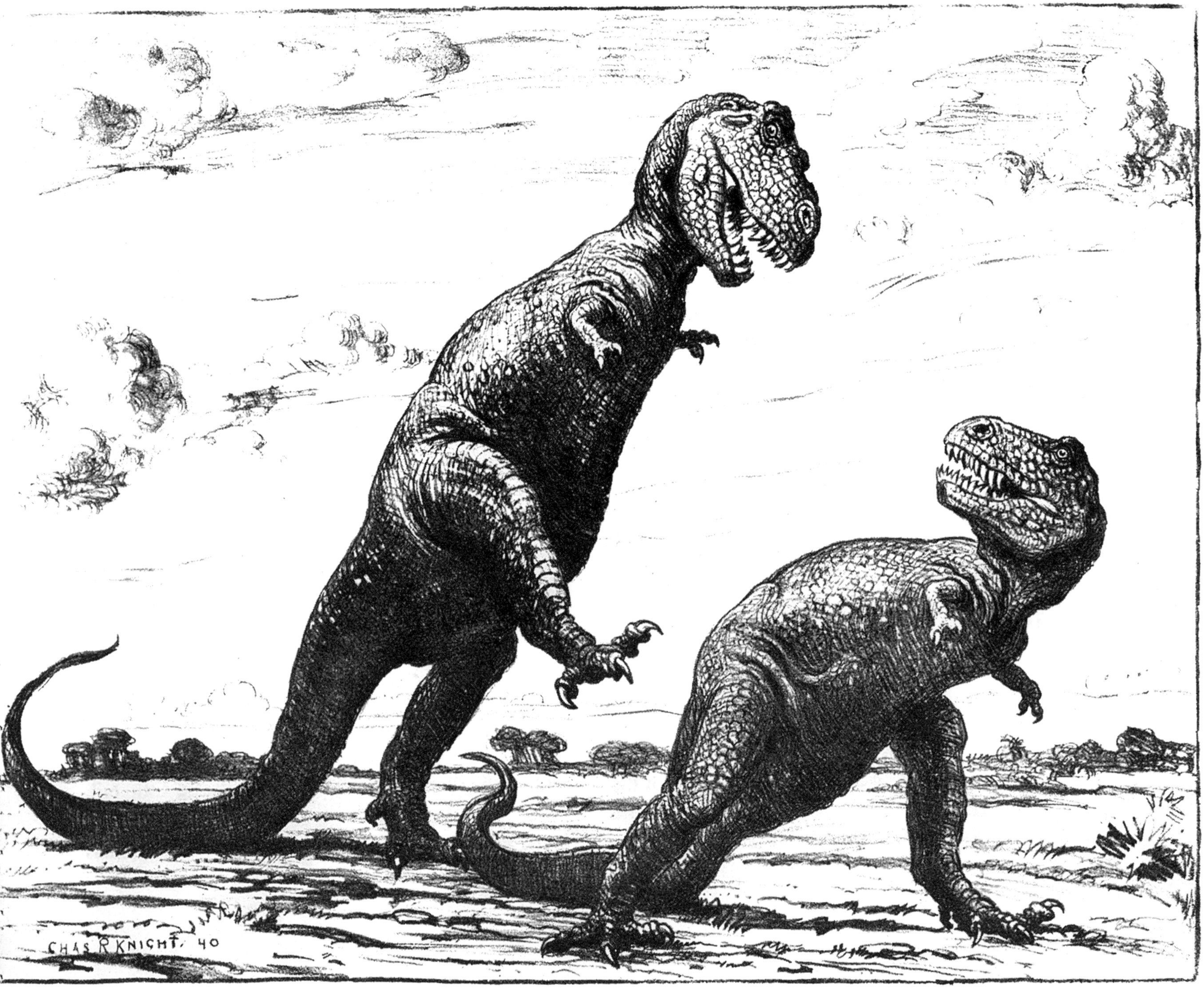
CHAS R KNIGHT '40

Birds

As far back as Cretaceous time, there were birds with small teeth in their jaws. This characteristic is a leftover from their Reptile ancestors. One fossil species in particular, a large loon-like creature, totally unable to fly, was especially adapted to an aquatic existence. The enormous webbed feet were set so far back that it was impossible for this expert swimmer to sit up. It doubtless rested upon its stomach on some sandy beach or convenient rock, whenever it chose to come out of the water. There were smaller flying types, however, quite like our modern terns. Later, huge flightless land forms appeared, among them a giant eagle-like, long-legged species from South America. The head of this creature was enormous, as big as that of a horse, but its tiny wings precluded any idea of flight. So far as we know there were no beautiful feathered beings in our early world. Our actual knowledge of fossil bird forms is scant, for the fragile skeletons of these flying creatures were easily destroyed. As we approach more closely to our own time, we know of numerous species—discovered in various parts of the world—of which the great moa from New Zealand is an important example. These birds were some ten feet in height. They no doubt existed in large numbers on the islands of the Pacific until about six hundred years ago, when their paradise was rudely shattered by the advent of certain South Sea natives. These promptly killed and ate the defenseless, wingless, thick-legged creatures, whose only safety lay in getting away as quickly as possible from their dreaded enemies. Today we know of them only by the discovery of an occasional fossil egg and a few scattered skeletons and feathers. In general appearance they resembled the ostrich. Their short, heavy feet carried four instead of two toes, they had no visible wings and the head was tiny.

Another species known to us is the dodo, a ground-living relative of the pigeons, which was destroyed in like manner by Portuguese and Dutch sailors, who landed on the island of Mauritius in the fifteenth century. We know the actual appearance of this extraordinary bird because of a picture painted from the bird itself by a certain Dutch bird painter. It shows the fifty-pound ball of fat flesh and flimsy little feathers, with a huge head and beak, and an absurd little tail.

Still another interesting though now extinct bird is the great Auk, found at one time on certain islands in the North Atlantic.

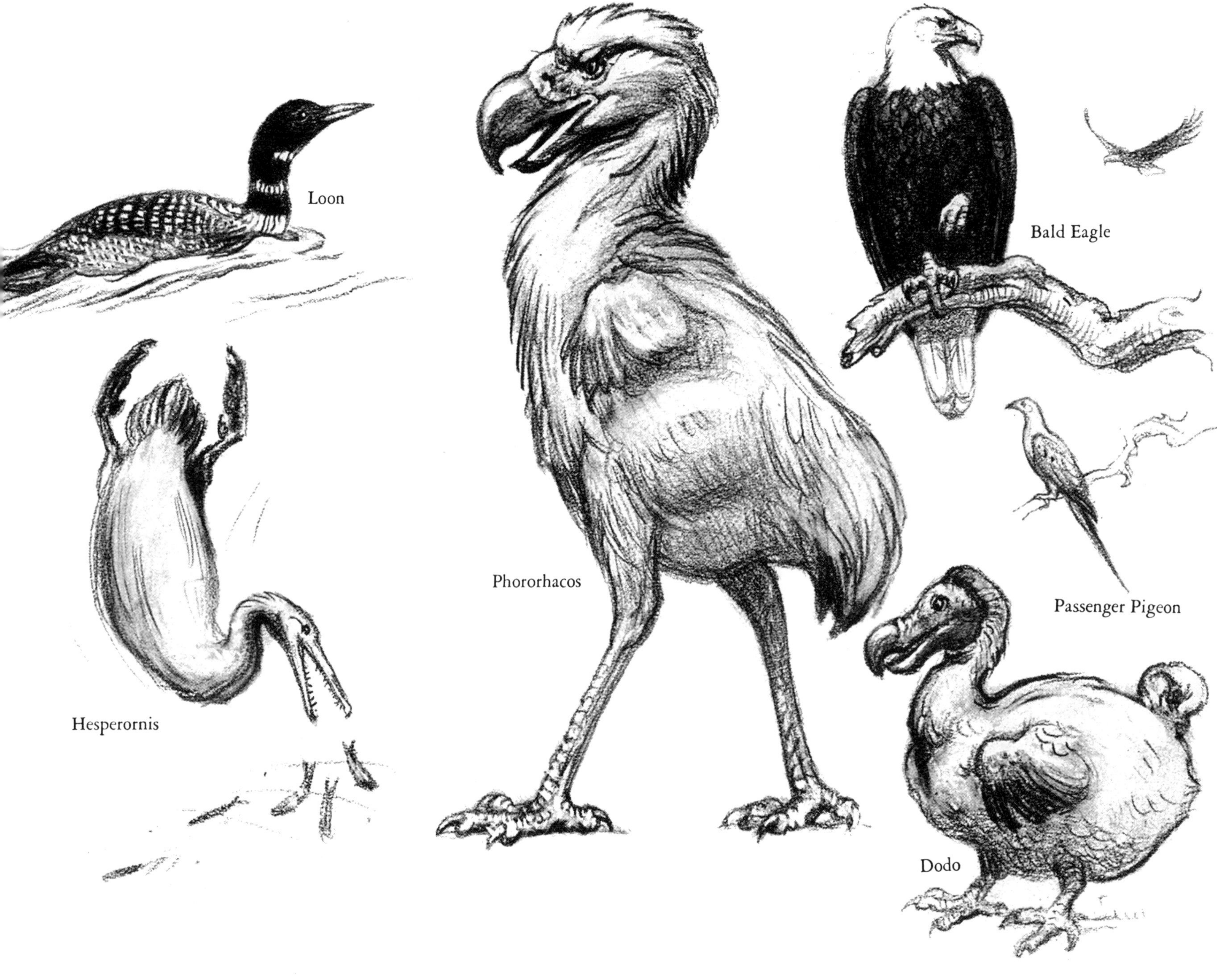
Loon
Bald Eagle
Phororhacos
Passenger Pigeon
Hesperornis
Dodo

Sharks

In shark-infested waters the cry, "Man overboard!" sends a thrill of terror to the heart of every mariner. At such a time the entire crew of the vessel leaps into action—boats are lowered and the men row frantically toward their unfortunate comrade who may possibly be rescued without injury from the jaws of the great fish. Fear of the powerful creature as it cuts swiftly through the deep blue water is ingrained in most sailors, particularly those who have seen, at one time or another, how quickly the dark misty shapes gather about the body of a dead whale, tearing huge pieces of flesh from the floating carcass, or perhaps impaling themselves on the sharp pointed shark-hooks baited with pieces of salt pork. While not all species will attack man, the great white shark figured in the drawing will undoubtedly do so whenever occasion offers. The panel at the right shows the life-size drawing of two teeth, one vastly larger than the other. The huge triangular cutting weapon belonged to an immense prehistoric species; the smaller was taken from the jaws of a thirty-six-foot white shark, a more recent species. The fossil was a colossal brute probably sixty or eighty feet long, with an enormous mouth capable of swallowing whole an animal as large as a cow. This monster is known only by his teeth, for the skeleton of a shark is cartilaginous and seldom preserved. One look at the mammoth tooth is enough to give one the creeps. Very fortunately this shark no longer exists, though our man-eating species, even when comparatively small, is a very dangerous creature. Today, as always, the oceans of the world are full of sharks of many different kinds—great whale sharks with big mouths and tiny teeth; hammer-heads, their eyes at the end of long stalks growing from either side of the head; spotted tiger sharks; pointed-nosed mackerel sharks; and dogfish, in vast numbers. They are all carnivorous. The dogfish devour quantities of lobsters off our coasts; the other species are mostly fish eaters. They are all graceful, swift, voracious, stupid, and very tenacious of life—a combination which spells destruction to the creatures on which they prey. Their skin is rough, like sandpaper, and dry looking, unlike any other fish that swims the sea. Tradition has it that a shark must turn over to seize his food. This is a mistaken idea, though a very general one, even among fishermen who should know the real facts owing to their constant association with the animal in its natural environment.

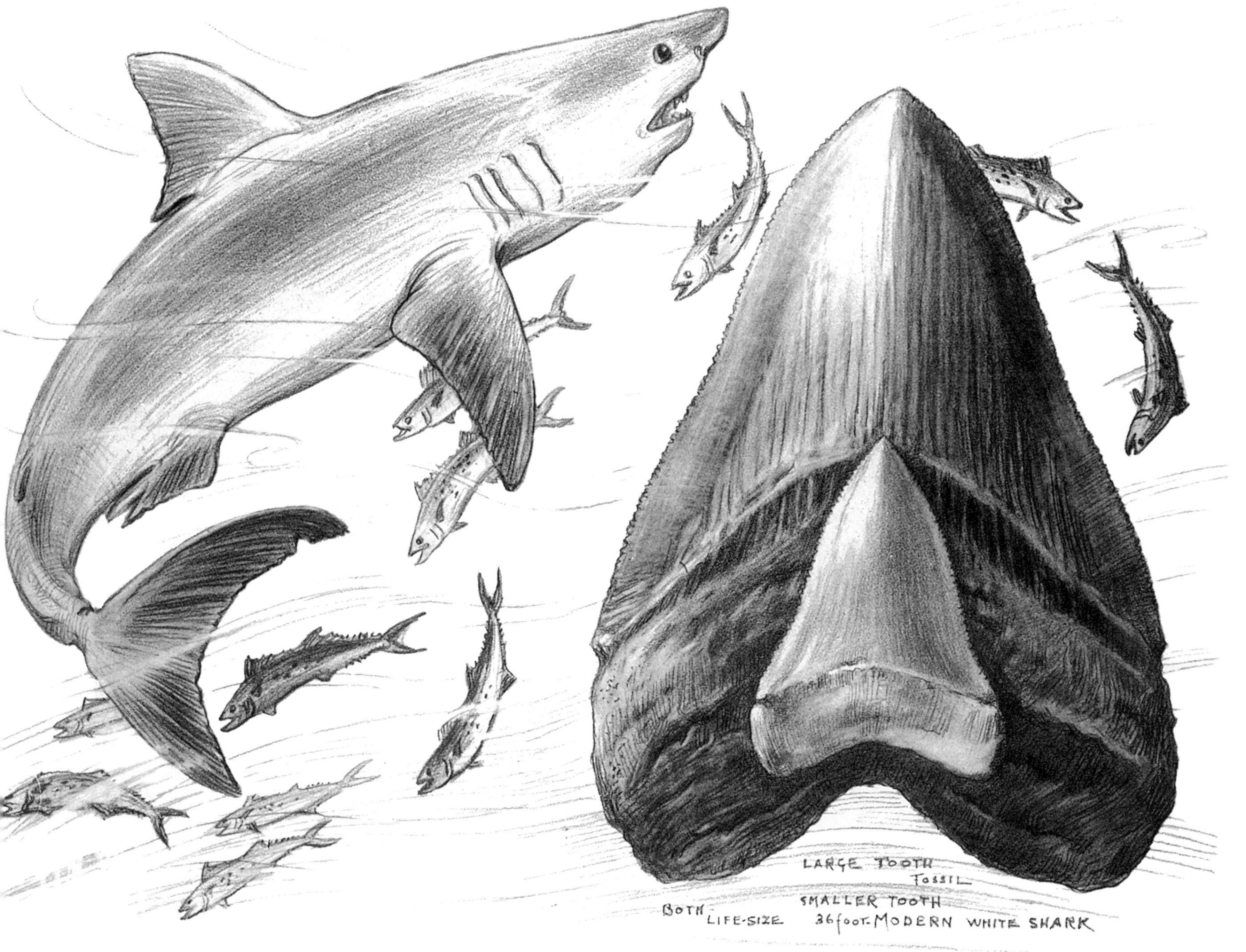
LARGE TOOTH
FOSSIL
SMALLER TOOTH
BOTH LIFE-SIZE
36 foot MODERN WHITE SHARK

Zeuglodon

To judge by the tigerish grin on the long jaws of this great sea-monster, life must have been a grim joke in his day and time. Food was plentiful and there were quantities of large glistening fish to be devoured and warm clear water to swim about in. He belonged to a class of mammals, the whales or Cetacea, as they are called, which scientists believe returned to the water after having been developed through countless ages upon land. Owing to their altered environment, the whale family lost the use of their front limbs which gradually evolved into paddles. The hind legs vanished almost completely. The tail, which grew to great proportions and was flattened horizontally, served to assist the animal in diving and propelling itself through the water. Remains of Zeuglodon have been discovered in some of our southern states. Today whales are the largest of all living creatures, even perhaps the greatest that ever existed. There are many species of whales, among them the one-hundred-foot Sulphur-Bottom Whale and the Right Whale. Others of the Baleen type are being destroyed for their valuable oil and other products. Baleen is the name given to the peculiar horny substance resembling a dense fringe, inside the rim of the upper jaw of the whale, through which the creature squeezes out great mouthfuls of water, retaining the small marine animals on which it feeds.

Sperm whales have huge blunt heads, their long thin lower jaws lined with rows of pointed peglike teeth. They feed on the giant squids, crushing the soft tentacled monsters and swallowing them in huge chunks. Spermaceti, an extremely valuable oil, is ob tained from the head case or skull of the animal.

The small whale, known as the Orca, Grampus, or Killer, is the terror of the northern seas. It swallows sea lions whole, tears out the tongues of the big whales, and even catches and destroys porpoises. A six-foot fin rises straight up from its back and the sleek body is banded and streaked in black and white.

There is also a white whale and a singular creature, the narwhal, with a very long, twisted, ivory tusk or tooth projecting straight forward from the tip of the nose.

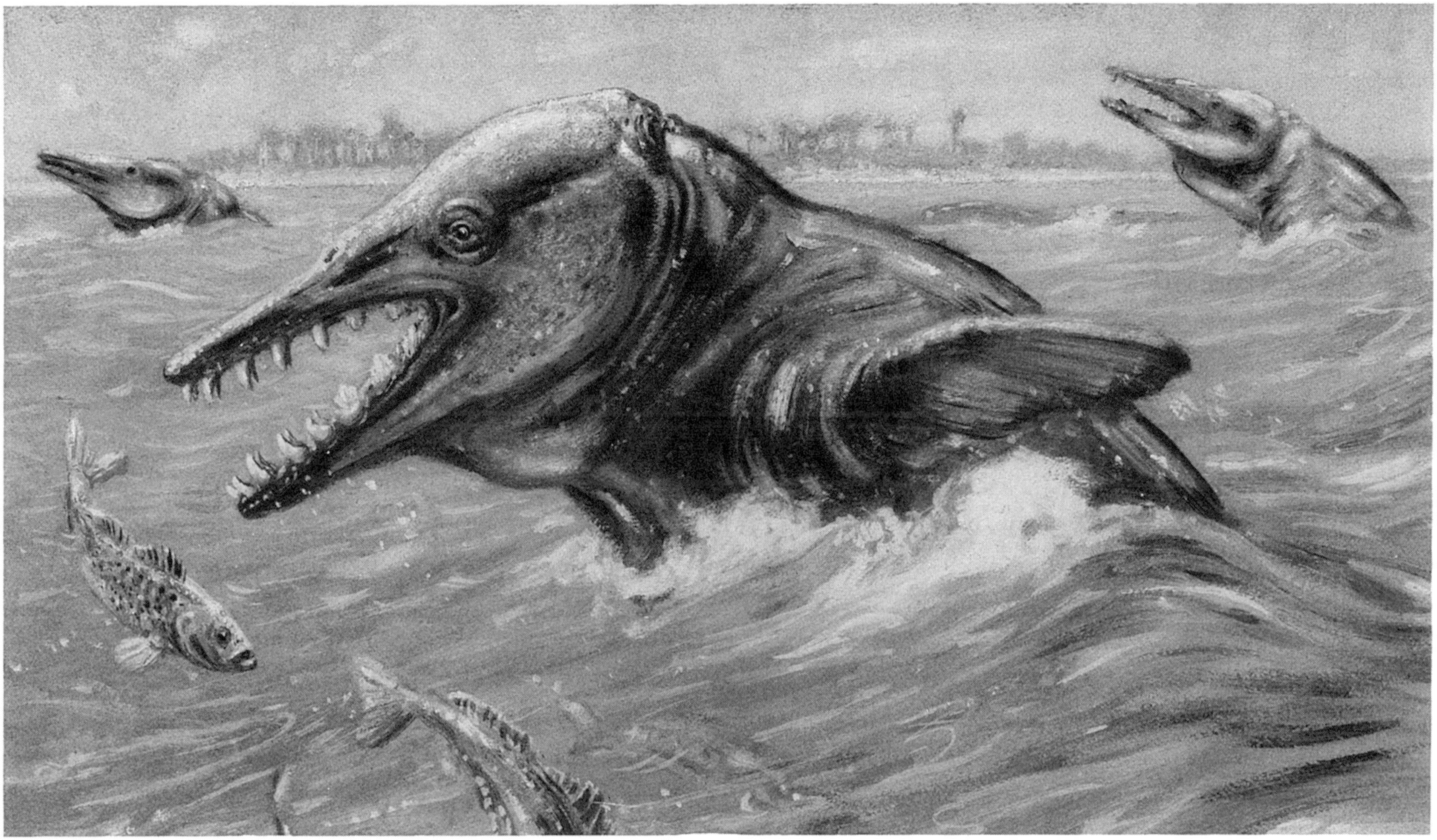

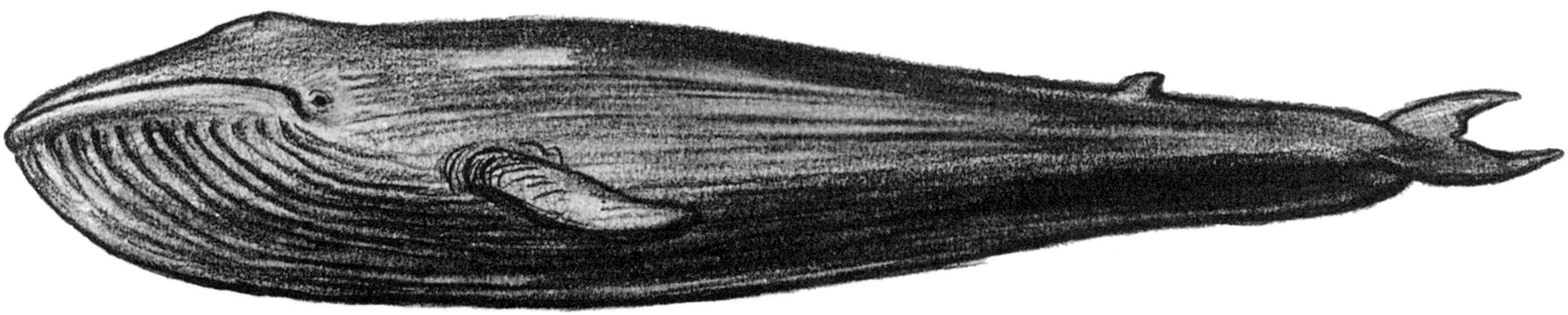

Later Reptiles

We have taken a peep at reptiles of the past but here we see a few modern types which still have some of the characteristics of the earlier ones. Our modern alligator could easily have come straight from ancient Jurassic days. The Galapagos tortoise, too, gives one the impression of venerable and respectable antiquity, but the great Komodo lizard is just a large species of the Monitors, a tribe of semi-aquatic reptiles now living in India, Australia, and other eastern countries. I drew this big fellow from life as he strolled about in the New York Zoological Park just after arriving from Komodo Island, near Bali in the South Seas. Quite recently discovered, these fierce and courageous reptiles are the largest living examples of the family. Some of them are as much as ten or twelve feet in length. They have sharp teeth and big recurved claws. When they were captured, they put up a terrific fight. But the largest one escaped, tearing his way out of the trap by sheer strength. The strangely grotesque head of the rhinoceros iguana, a four-foot species found in certain West Indian islands, is reminiscent of the stone image of some Aztec god, with its points and horns and heavy hanging throat pouch. The armor, like that of so many other animals, is purely for defense, as the creature, although it will scratch and bite if molested, is quite harmless. Even the tiny so-called horned-toad of our western deserts is really a lizard and were he enlarged to the size of a great dinosaur would be even more terrible in appearance. The long-necked, heavy-shelled tortoise is a species from one of the Galapagos Islands off the coast of Chile and is probably an old individual. We think, indeed, that these hoary specimens really reach a greater age than any other living creatures (perhaps two hundred years).

Coiled to strike, the big western diamond back rattler sounds his warning note and turns his flattened head and glittering eyes towards the intruder. The long sharp fangs which lie against his upper jaw will rise into fighting position when the mouth is opened. Death lurks in the coiled body of the glittering reptile.

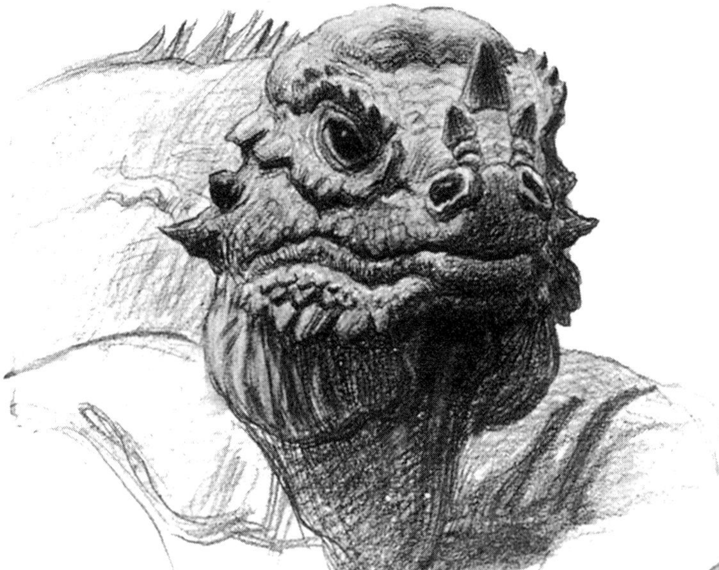

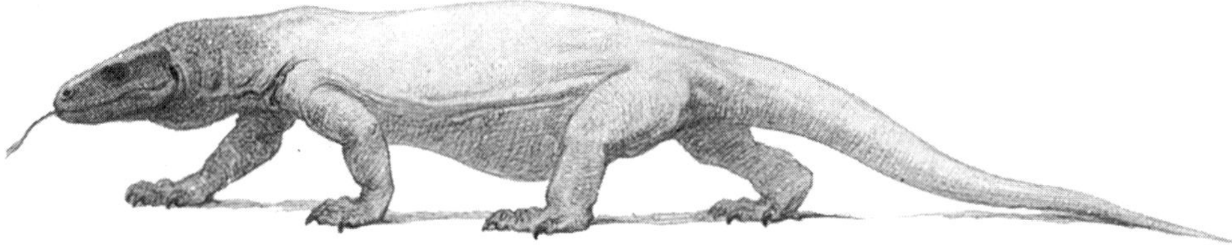

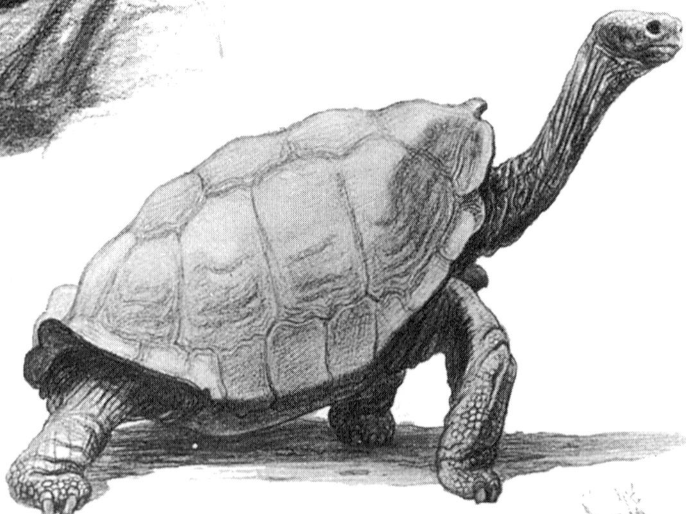

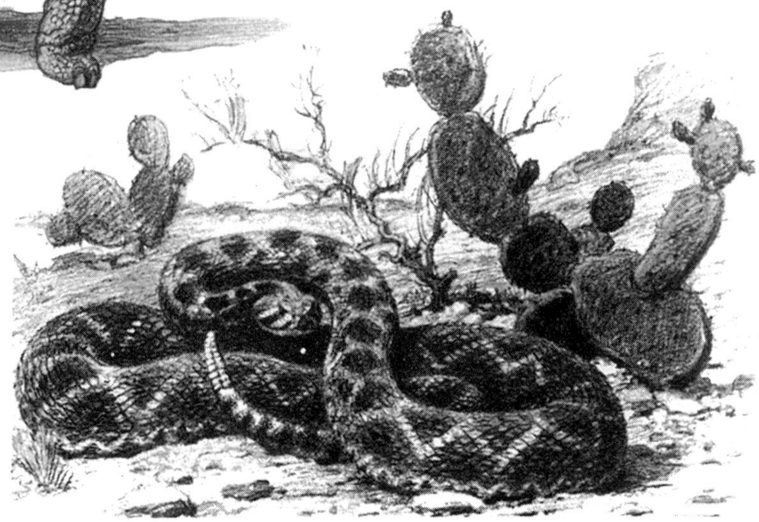

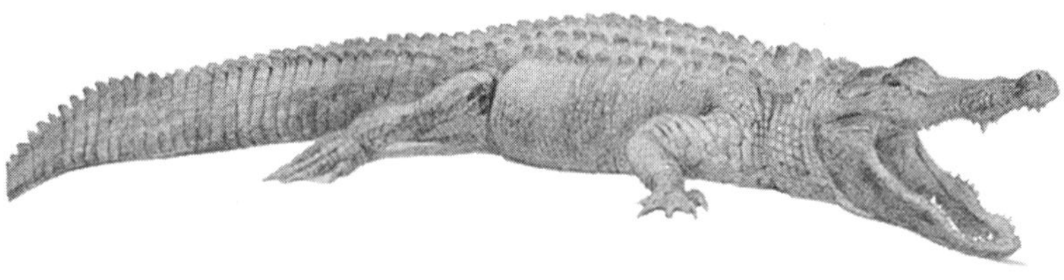

Early Elephants—Eocene Period

The story of the evolution of mammals is very complete in the elephant family. During various eras certain types of these interesting animals ranged far and wide over most of the continents. In this country, they have a very long history—the earlier four-tusked, long-jawed species—some large, some small—slowly being replaced by the later short-faced forms. It was in Egypt, however, in a region south of Cairo called the Fayum desert, that the earliest examples were discovered. Skeletons of these animals, together with various water species, excavated in the region of what was once a lake (Moeris), proved a wonderful discovery. The two extremely early elephant-like varieties —one hardly yet an elephant, though tending in that direction—the so-called Moeritherium, and the other, the perfect little four-tusked Mastodon, showing all the characteristics scientists had hoped to find, were rich treasures for the expedition sent out to hunt for them. It may seem strange that all early elephants apparently had four incisor, not canine, tusks—two in the upper jaw curving slightly downward—and two in the lower jaw growing out to meet them. The lower jaw was very long and slim, not at all like that of our modern elephant with its shortened stumpy face. Elephants of a more recent period have lost the lower tusks completely. The upper tusks have developed enormously, especially in the mammoths and in our great modern African elephants. Originally, the little Fayum species and some of the later kinds, had long bodies, short legs, and very small trunks. As time went on, however, the elephant grew taller and taller and his trunk longer, too, in order that it might still touch the ground. The mammoths and the two species of elephants in existence today have the long snout (which is really the drawing out of the nose and the upper lip) very much developed. Without this unique trunk, the elephant would soon die, for it uses it to breathe through, and to help it drink. This it does by drawing the water up into the trunk and then, inserting the tip into its mouth, it blows the fluid down its throat. But most important of all, the powerful twisting proboscis collects the animal's food—hay, leaves, anything it chooses to eat —and draws it to its mouth where it is well chewed before it is swallowed. The illustration shows a progression in size and form of both the trunk and the tusks, as well as the body of these most intelligent pachyderms, through millions of years of their existence.

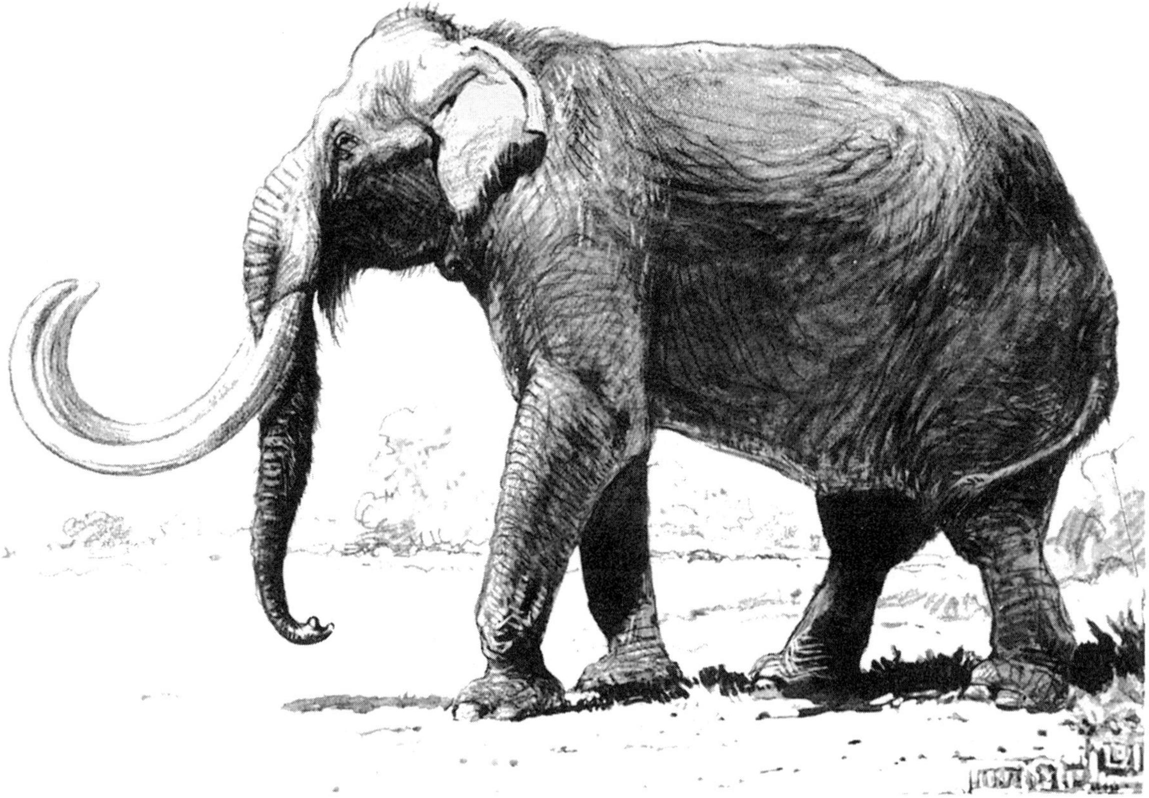

Mammoth

Moeritherium

Early Mastodon

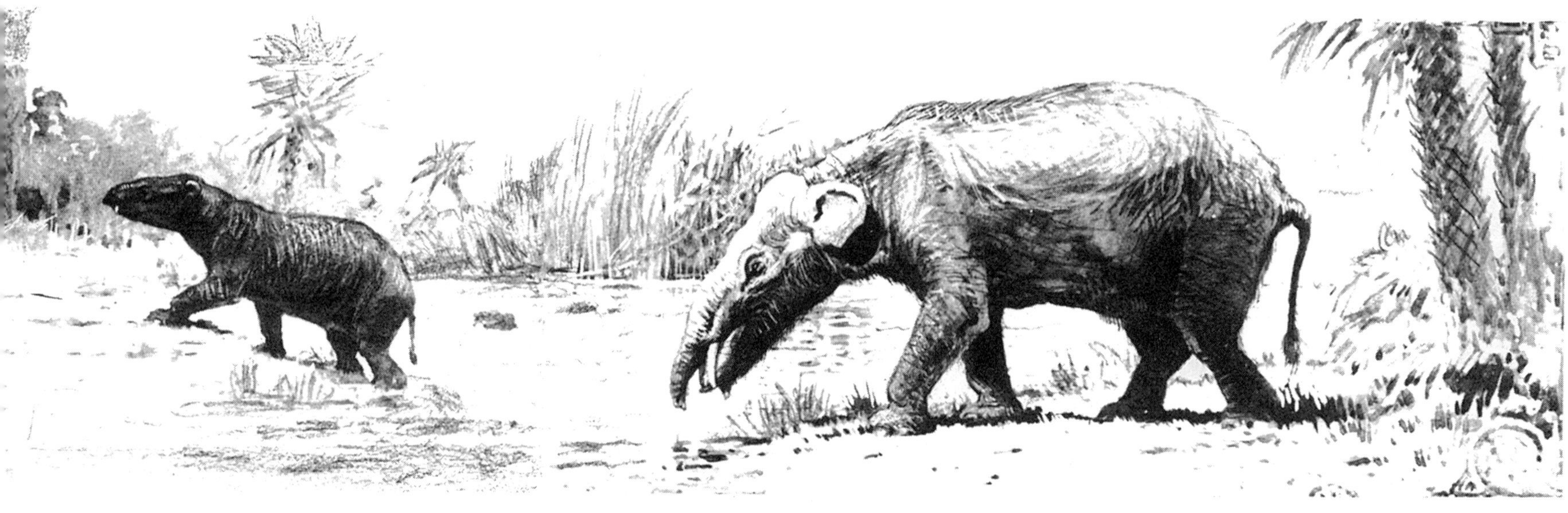

Courtesy American Museum of Natural History

The Woolly Mammoth—Pleistocene Era

Once the world must have been full of mammoths to judge by all the huge curling tusks, great leg bones, and bits of flesh and fur that one finds in such widely-separated places as Siberia and Florida, Alaska and France. Their relatives, the mastodons—squattier and more clumsy creatures—were distributed over many sections of the United States, New York State in particular being one of their favorite habitats. Whether early American man ever saw a mastodon is doubtful, but in Europe, at any rate, our primitive ancestors did see and fight with, and at times eat the great Woolly Mammoth. Even I have tasted mammoth bone marrow, a bit from Alaska, where many specimens of these monstrous animals have been found, still quite well preserved, with skin and flesh adhering to the bones. But the most wonderful examples of these great creatures, frozen almost whole in the ice, have turned up in Siberia. They were covered with thick reddish wool and hair, had small ears, a very short tail, and enormous, much-recurved tusks. Mammoth meat, probably because there was so much of it, was eaten with gusto by our ancestors in European countries, and at certain spots in the Balkan region remains of many young ones have been found, the long bones cracked for the marrow. Why these sagacious animals should have found it necessary to brave the inclement weather of a winter in northern Europe is always a great puzzle to scientists, but evidently they didn't mind the cold and food was plentiful. Our illustration shows early hunters trying to hamstring a mired mammoth by jabbing it fiercely with a flint-tipped spear. Only desperate hunger could force these puny humans into such a precarious position. One of them is evidently done for as he flounders helplessly at the feet of the huge animal towering above him. Life in those days was a strenuous affair, for both man and beast. A constant feud went on between them, man in most cases being the aggressor. Gradually, by virtue of superior brain power, he gained ascendancy over all the creatures of the field.

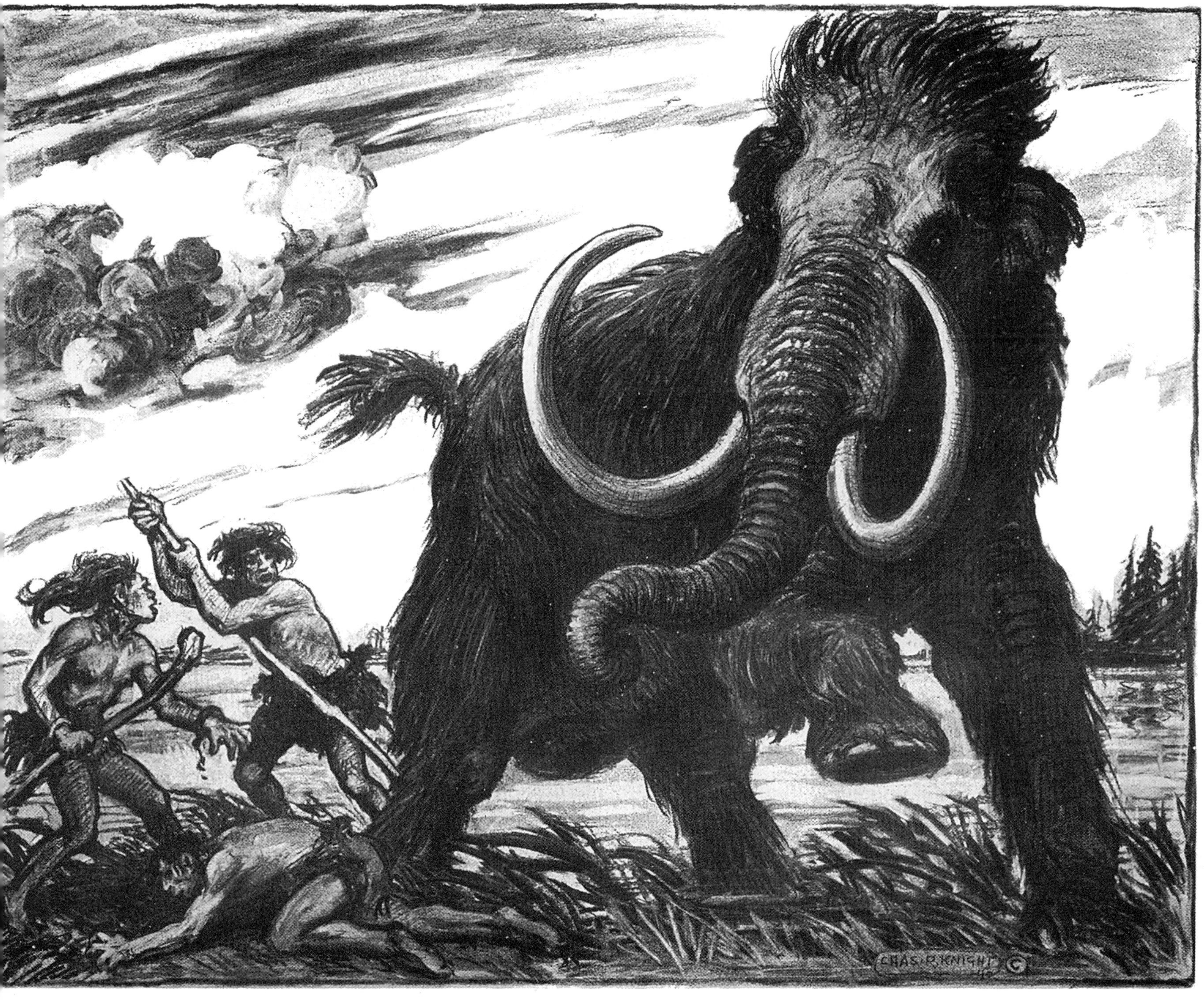
CHAS. R. KNIGHT ©

Modern Elephants

In stature, neither of the two magnificent species of modern elephants, the African and the Indian, are quite as large as the very biggest of the mammoths but to us who have never seen a mammoth they certainly seem so. By actual measurement our African species may top eleven feet, though they do not often come as tall as that—while ten feet is well above the average for the Indian variety. "Jewel," the Indian beauty shown, is a circus queen, but being a lady she has no visible tusks, whereas in the African variety both sexes carry them. Large tusks are scarce today, most of the finest of these tusk-bearing animals having been killed off long ago. I have purposely drawn both species from the same general angle so we can compare them easily. The top of Jewel's head is evidently much higher and rounder than that of the African species, the ears very much smaller, and the trunk longer and more delicate in form. Though she is not so tall she is distinctly plumper and rounder, with thick legs, while he has very long, slim legs, gigantic ears and a trunk that is thick at the base. Reams could be written about the habits and peculiarities of these two great creatures and tall stories are rife, particularly about the intelligence of the Indian species. There is no doubt, however, that elephants are exceedingly canny and that they learn easily and have excellent memories. A glance at Jewel will show that her brain power is of a high order, and to see her handling her great bulk so easily, obeying orders, keeping out of the way of passing objects, and altogether conducting herself in a thoroughly lady-like manner, is really a most entertaining experience. African elephants are not supposed to be very bright, although of late years they have been trained in many ways, though not quite so successfully as their Indian cousins. Yet it shows what can be done along those lines. Surely no grander creature ever lived than the male African elephant, immense, post-legged, practically hairless, with great flapping ears. Standing in the shade of a spreading acacia tree, the huge gray form seems to typify everything sublime in a world of animal life that is destined all too soon to pass away forever.

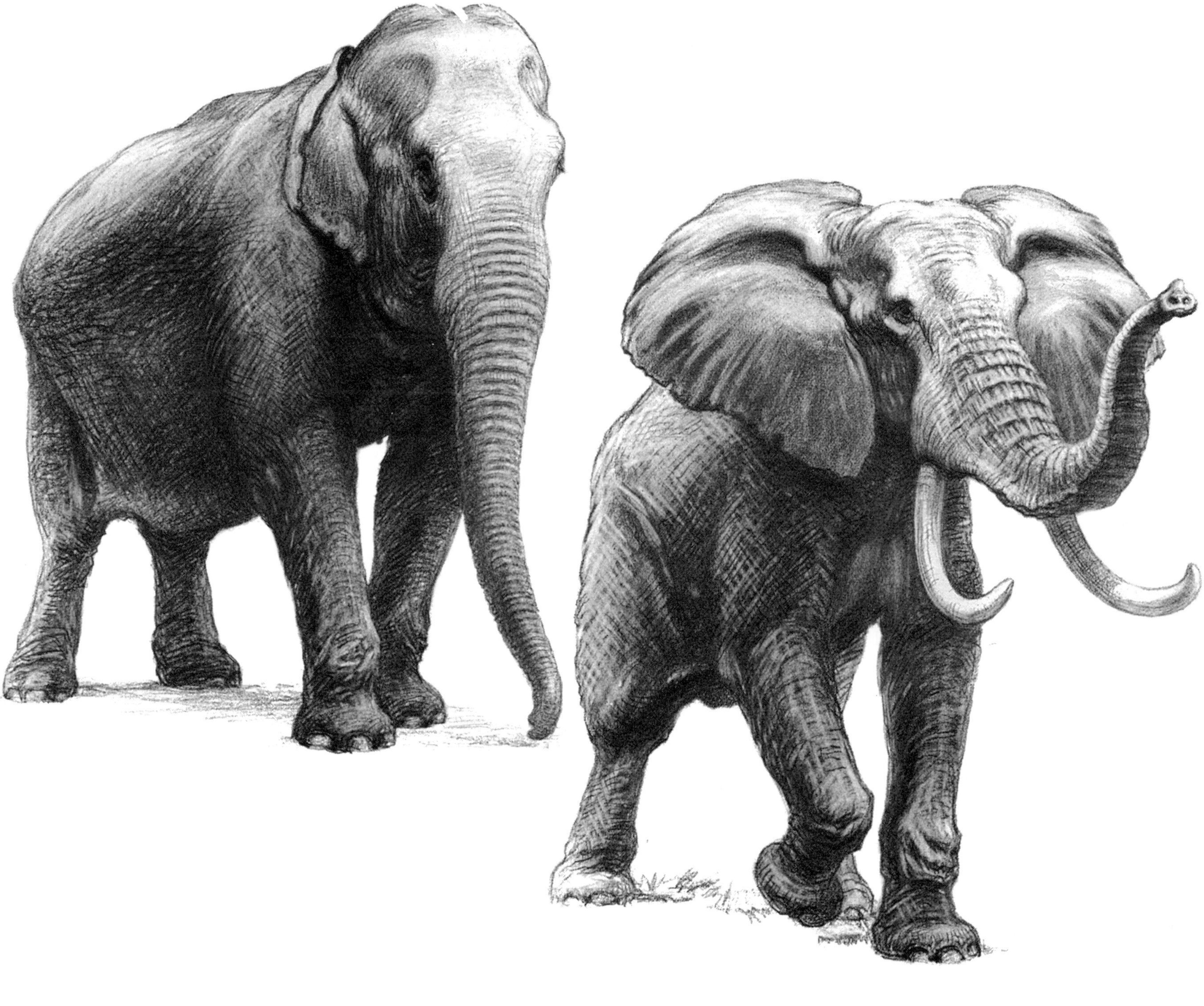

Dogs—Fossil—Modern

Surrounded as we are today by so many types of canines, both large and small, we hardly realize that every one of them has been artificially bred by man and that no such varieties existed in a natural state. The few wild species are most of them of the wolf, fox, jackal type, with upright pointed ears and bushy tails. Even the origin of our domestic breeds is shrouded in mystery, some of them, notably greyhounds, being portrayed in very early paintings by Egyptian artists. The Assyrian reliefs also show many dogs of the large, drooping-eared mastiff type, a breed which had come a long way from its ancestral relatives. We do have certain wild forms which would hardly be recognized as dogs at all. They are small creatures technically canine animals, somewhat resembling a racoon—offshoots from the main stock. Early fossil dogs do not closely resemble modern species. Like the big Daphoenus in the drawing, they are more the proportion and size of a leopard, with small heads, curved backs and very long, drooping tails. In my opinion, our big recent civets would not be unlike these early canines in general appearance. The Egyptian pictures also show a pointer and many small curly-tailed dogs; and occasionally, a tame jackal running with the pack. Our domestic chows, collies, Alsatians, Samoyeds and Esquimaux dogs do look, more or less, like wolves, though careful breeding has greatly altered the chow's appearance in the last few years. Such breeds as bulldogs, Pekinese, dachshunds, Great Danes, etc., are either giants or dwarfs, as the case may be. They would not last long in a wild state, as they depart too greatly from the normal dog size, which is about as large as a small Alsatian. The wild dog of India, called the dhole, is perhaps nearest to our domestic strains, though he looks more like a fox than a dog. He hunts in packs, however, and is a persistent and voracious killer of wild game. It would seem to me that some animal not unlike the dhole, or better yet, the Australian dingo, might have been at least one of the ancestral types of our domestic breeds. At any rate, when a number of our homebred varieties are allowed to mix freely, the offspring will be, as a rule, an upright-eared, medium-sized "yellow dog."

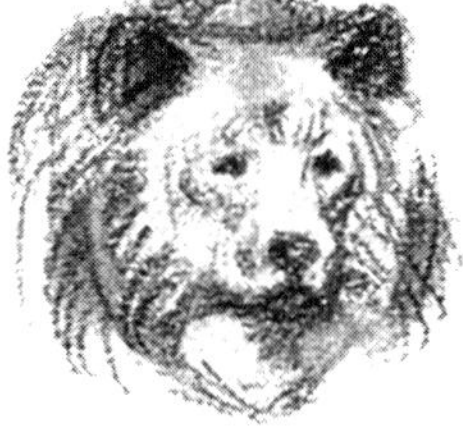

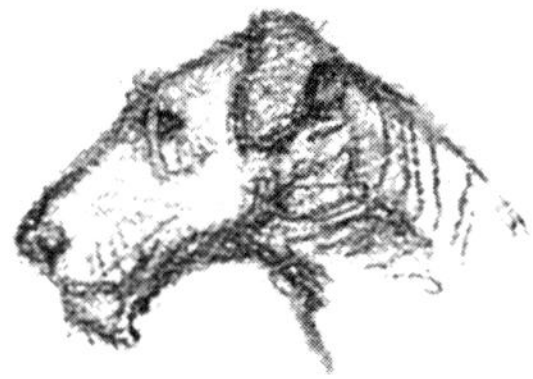

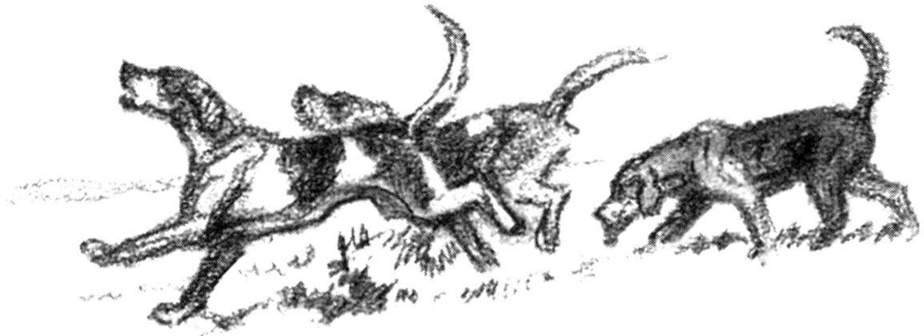

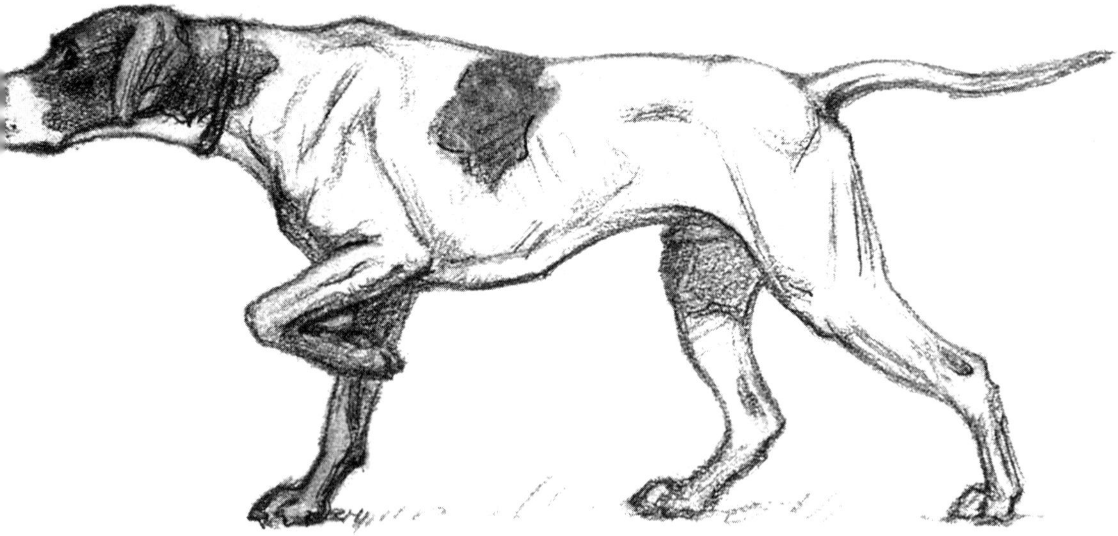

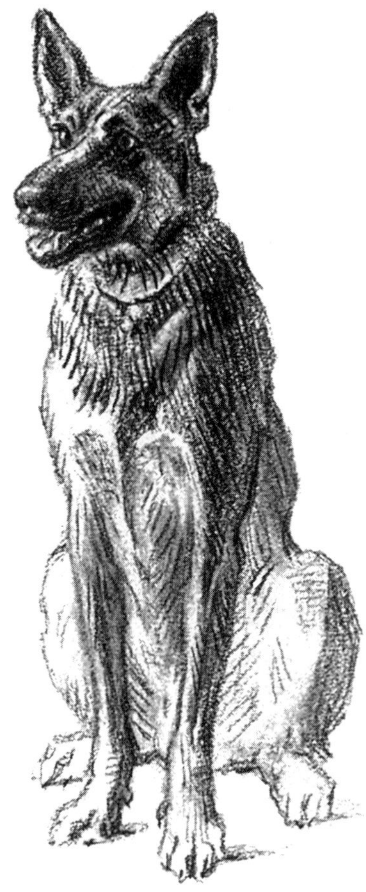

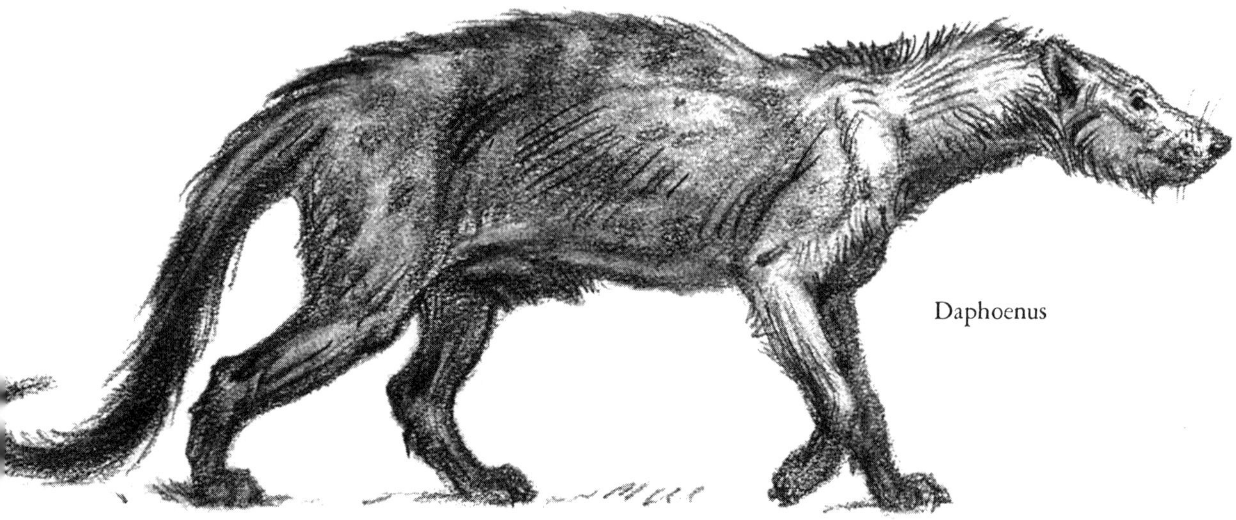

Daphoenus

Cats

If anyone were to ask me how I would separate the cat family according to their mental reactions my reply would be, "Put the lion at one side and all the others, including the tiger, in an opposite category." This is not a haphazard opinion but one that has been borne in upon me after many years of drawing and painting these magnificent beasts. Structurally, indeed, they are all singularly alike—only an expert can tell a lion skeleton from that of a tiger. The king of beasts is a thing apart in many ways. His regal mien and general upstanding carriage are well shown in the fine photographs one sees nowadays, taken right on his native heath in the heart of Africa. Here the great beast is at home and one notices immediately that several individuals (sometimes ten or fifteen) go about together, sleeping, feeding, sparring, hunting in a herd or "pride," as the old English chroniclers have it. How bold and indifferent their attitude to man; while tigers, leopards, jaguars, pumas—all the rest of the tribe—remain aloof, suspicious and savage. Cats, in early times, seem to have favored the sabre-toothed type, which finally took the form of the great so-called tiger. He was a mighty feline with an enormous gape, long thin sabre-like teeth, a powerful body, thick legs, and huge clawed feet. The jaws that opened wide were necessary to the animal's existence, as only in this way could the points of the long fangs be brought into play as they struck downwards into the victim. They were like the fangs of a rattlesnake, cutting veins and arteries, and bleeding the hapless prey to death. All modern cats seize their prey with open mouth, driving both upper and lower canines into the animal attacked, in a manner quite different from that of their big predecessors. As killers, cats are very successful, creeping upon their quarry and striving to overcome it in one fierce rush. If, however, the animal should succeed in breaking away, cats will not, as a rule, pursue it, knowing well their own limitations as to speed and endurance. I have compared a drawing of an actual tiger with the sabre-tooth, in order to show the very decided differences between them. By looking closely at both we see that the sabre-tooth was not a true tiger, but merely a superb early example of the feline race.

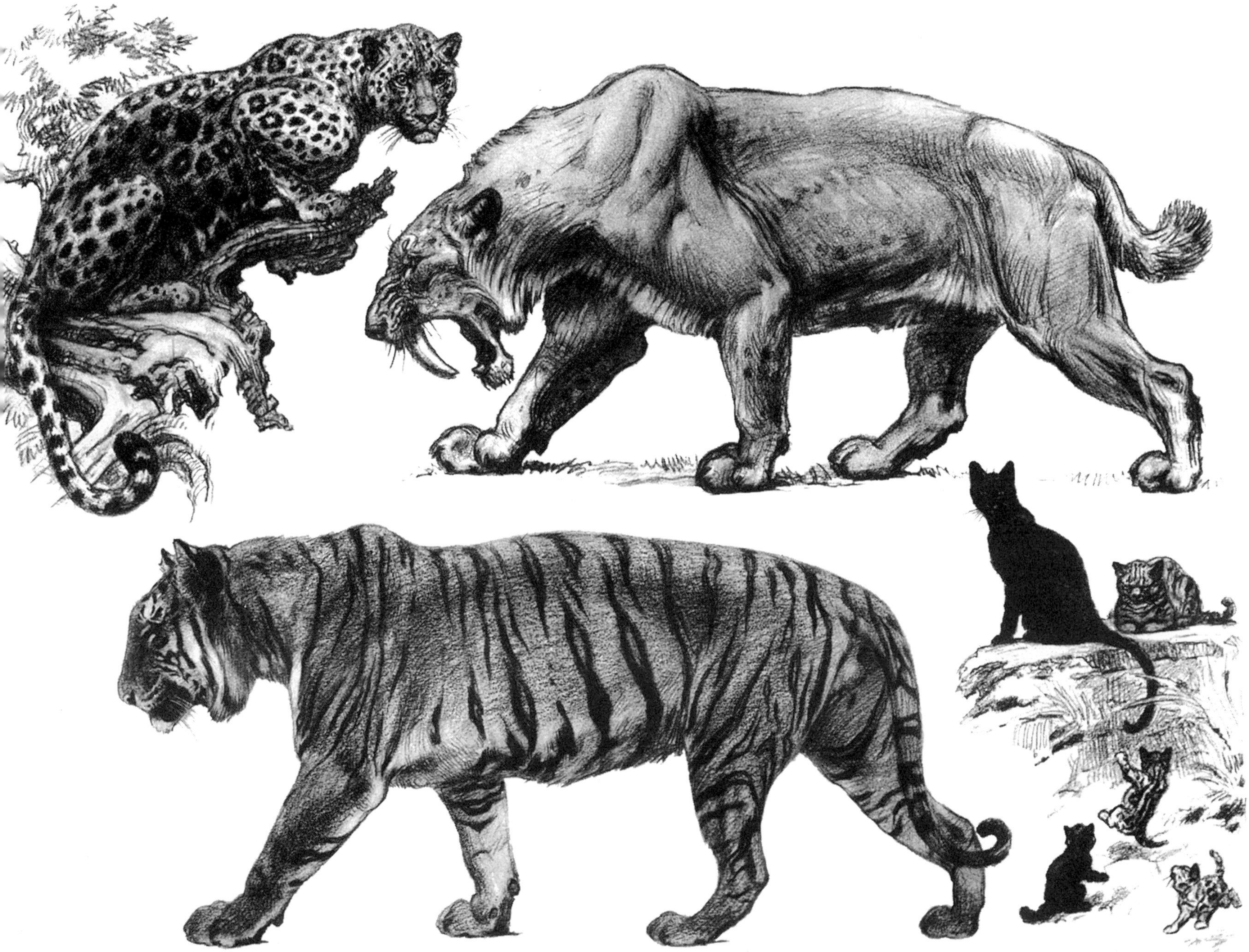

Rancho La Brea—California Pitch Pools

As a vein of gold looks good to a prospector, so does a cache of bones to a fossil hunter. When some years ago in Los Angeles, California, laborers, who were excavating an asphalt pit, came upon scattered bones, they reported their find to the proper authorities. The local scientists at once became intensely interested. Few, however, at first realized the extent or variety of the treasure, but as more and more bones were uncovered—remains of numerous individuals and many species—the importance of the discovery dawned upon them, and money and material were soon forthcoming to carry on the work. In order to protect the pits from encroachment, public-spirited men decided to turn this area into a city park. Hundreds, even thousands of skeletons have since been removed from the intensely sticky, sugary asphalt which surrounded them, and placed in the local museum. Peaceful as the place looks now, tragedy, both dark and terrible, long hung about its gloomy depths. It may be that water standing on top of the treacherous ooze attracted thirst-maddened victims to their doom, or that dust so covered the surface that it was invisible until it was too late to retreat. In any case, the pits were filled with the most widely-diversified types of early creatures—huge mammoths, mastodons, bison, horses, camels, ground sloths, wolves, sabre-toothed tigers, a big lion, and numerous species of birds. All these, and many more, had perished miserably at some time and had finally been engulfed and their bones disarticulated in the ever churning mass of asphalt. The scene before us depicts an incident of all too frequent occurrence in the dire history of the great death trap. A small camel has succumbed, exhausted by its futile struggles, and the usual motley crew of vultures, wolves and sabre-tooths has been attracted by the commotion. They do not seem apprehensive as to the fate which awaits any incautious moves on their part, but the hand of Fate reaches over them. Before tomorrow dawns at least some of their number will have perished in the black depths. Today, the viscous, dark fluid still oozes slowly from the vents, but no longer is the sinister trap baited with the body of some poor animal. Today, the "dragon" sleeps.

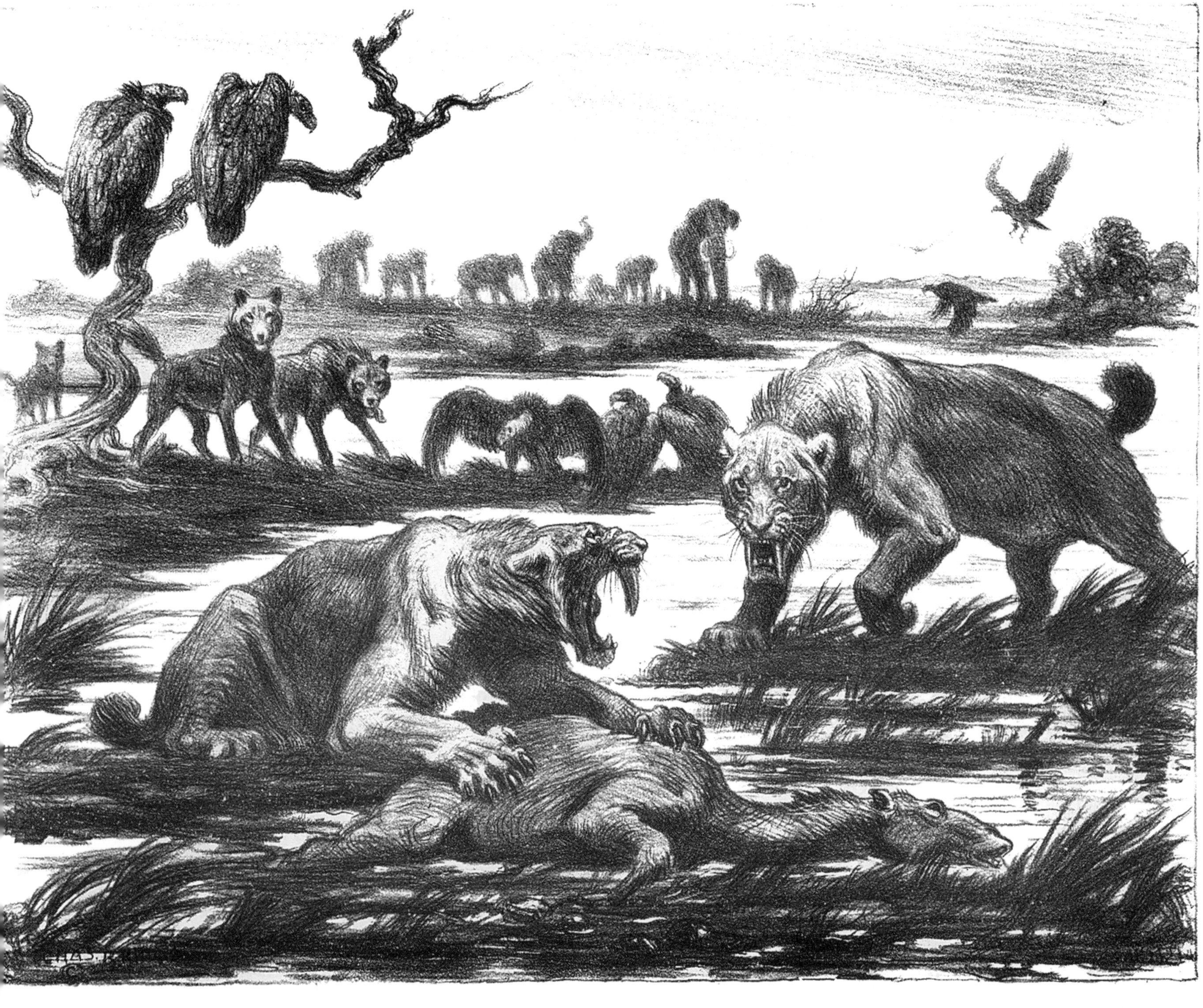

The Tiger

Ferocity incarnate aptly describes the terrible expression on the face of the brute before us now. From out of a maze of black and white and gold markings, cold greenish eyes stare balefully, and the huge mouth is opened wide to show the glistening fangs. Terrible, coughing roars issue from the capacious throat and the long whiskers around the mouth are thrown boldly forward. It is all terribly confusing for an animal painter, and it certainly takes a bit of fortitude to keep one's wits working and one's hand drawing to put down this fleeting expression under such difficult conditions. A peaceful tiger is one thing, but a tiger aroused is quite another. I purposely chose the tiger in this picture because in him are all the feline traits and it may be taken for granted that whatever he does other species will do under the same provocation. Like all cats, the lion included, the tiger is a nocturnal hunter and a dreaded destroyer of animal life. Deer, cattle, wild boars, and horses, even man himself, are all subject to attack. When once he acquires a taste for human flesh, the fierce creature becomes a real menace, oftentimes practically wiping out some miserable Indian community before he is laid low. The range of the big feline is extensive. He exists in the bitter cold of a Siberian winter, the warmer hill country, the hot marshlands of India, or the steaming jungles of Sumatra. Specimens shot in the northern section of his habitat are very large, and heavily furred, with much white showing on the underside. The farther south tigers are found, the shorter grows their hair, but they are still big fellows, brilliantly marked. The Sumatran animal is the smallest and darkest of his kind, being decidedly reddish brown in color with very little white visible. But wherever we find him he is the acme of all that is superb in the feline race, magnificent in color, statuesque and graceful in form, and lordly in general bearing. He stands as the grandest example of all the flesh eaters of the world.

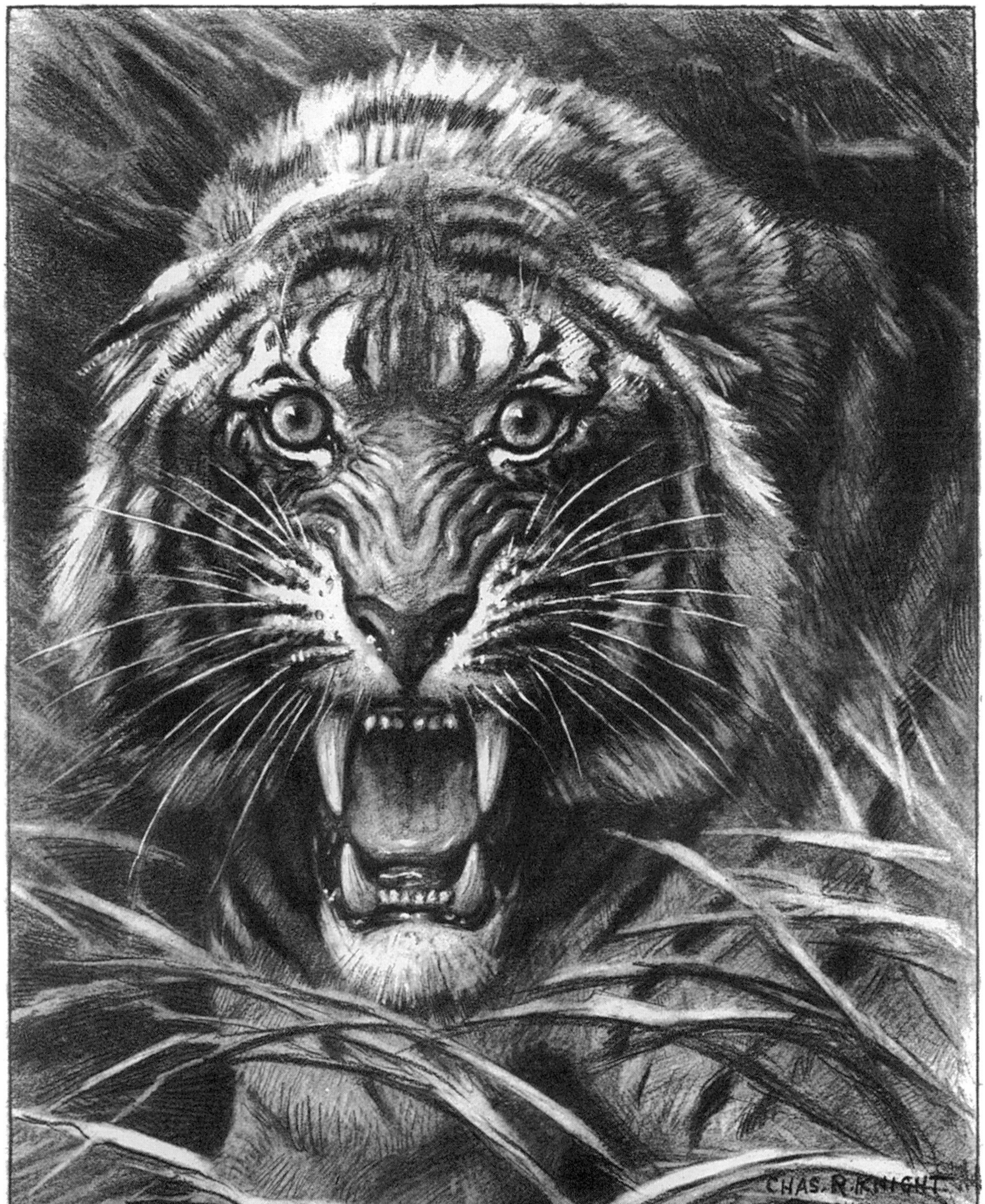
CHAS. R. KNIGHT.
©. 1943

Eocene Horses

The expert collectors of prehistoric animals, men employed by our great museums, are always looking for new and interesting specimens wherever they can be found. Many gaps naturally occur in all lines of fossil animals, but we have been singularly fortunate in coming across long series of prehistoric horses beginning back in Eocene time and continuing practically up to the present day. Little Eohippus (Dawn Horse), a very ancient and diminutive relative of our present animal, has always intrigued the general public because of his tiny form (twelve inches high) and his faraway connection with the beautiful creature we know so well. In his day, perhaps forty million years ago, changes were beginning to evolve in the world of mammals. Some species were destined to continue, others to disappear entirely in later ages, leaving no descendants. Our little Dawn Horse, light and fragile though he seemed, belonged to a fortunate type, for he was fast; he could jump like a flash and lose himself in a bush when pursued, or he could even run away from his enemies. As we look at him closely we notice something queer and different about his tiny feet, for unlike our modern horse, he has four toes, not one, on his front foot, and three, not one, on his hind foot. Each toe is covered by a stubby nail which isn't really a hoof as yet. In later ages, the four-toed type will gradually change to a three-toed form, larger and more graceful than its predecessor. In our pictures a little group of the diminutive creatures is running away from a big, slow flesh eater of the period. He will not be able to catch them, because they are faster than he. Indeed, very few animals then living could catch one of these horses, which is, no doubt, why they went on multiplying and increasing in size, strength and speed. We mustn't really think of this sprightly midget animal as a true horse, because at the time he bore little resemblance to his modern relative. The Dawn Horse was the earliest type of horse.

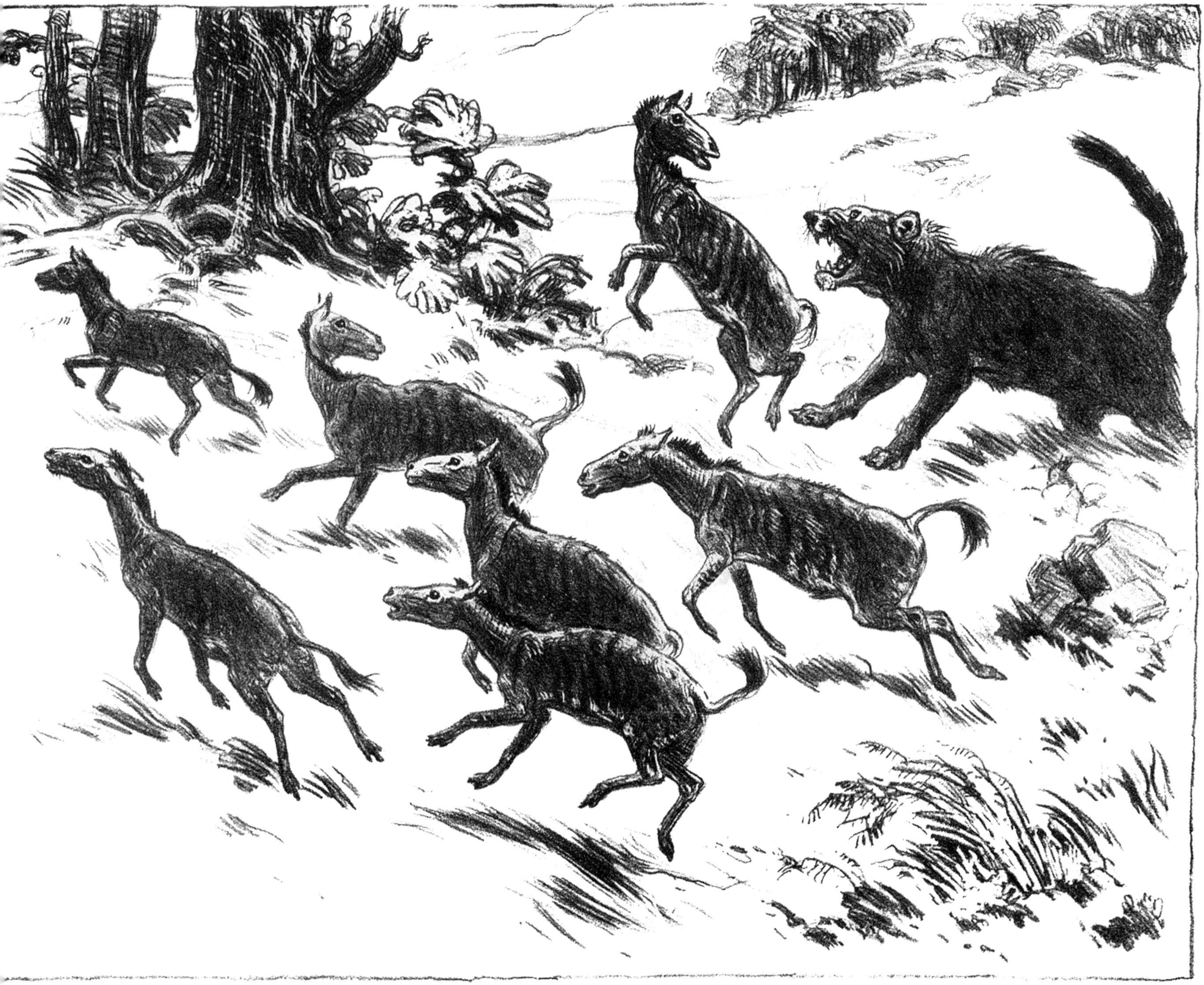

Hipparion

A very speedy little fellow is this three-toed ancestral horse, running so swiftly across the open glades. The delicate feet, each with its enlarged middle hoof are a distinct advantage to a fast-gaited animal, and indicate that the type has taken a long step forward structurally. As time goes on, the body grows more horse-like and even the head assumes an equine profile—long-nosed, long-faced, a deep, flat, smooth cheek, small ears, and probably a stiff little upright mane. The small side toes are evidently of not much use at this stage of the creature's development and will finally deteriorate into the so-called splint bones of the modern horse. This gradual doing away with unused parts is a law of nature which aims at keeping a creature at its maximum efficiency at all times. Speed was a characteristic of horses through many species and various ages. Their movement was so fast and sustained that they could outdistance any pursuing animal in a fair race. As long as some carnivore did not pounce upon them at a waterhole, the powerful little creatures stood a good chance of living out their natural lifespan. Anyone making a study of prehistoric horses is amazed at the number and the variety of the species. Some thickset, slow-moving types are supposed to have been forest dwellers, but in the main, and especially in later times, the horse was a creature of the open plains in many parts of the world, living in herds and grazing on the sweet and succulent grasses of the open prairie. This country in particular had a great number of them, some very large, others quite small. Finally, in the later stages of their development, they attained the true horse character—a toe covered by a hard rounded hoof. To the observer, a horse, a tapir, and a rhinoceros do not seem in any way related. Yet, if we go back far enough in the history of all three animals, we will be astonished to see that at certain periods of their existence there was not much difference between them. For in those days the rhinoceros was a small wiry little fellow with no nose horn; the tapir looked much as he does now; and the horse had not as yet developed any special characteristics.

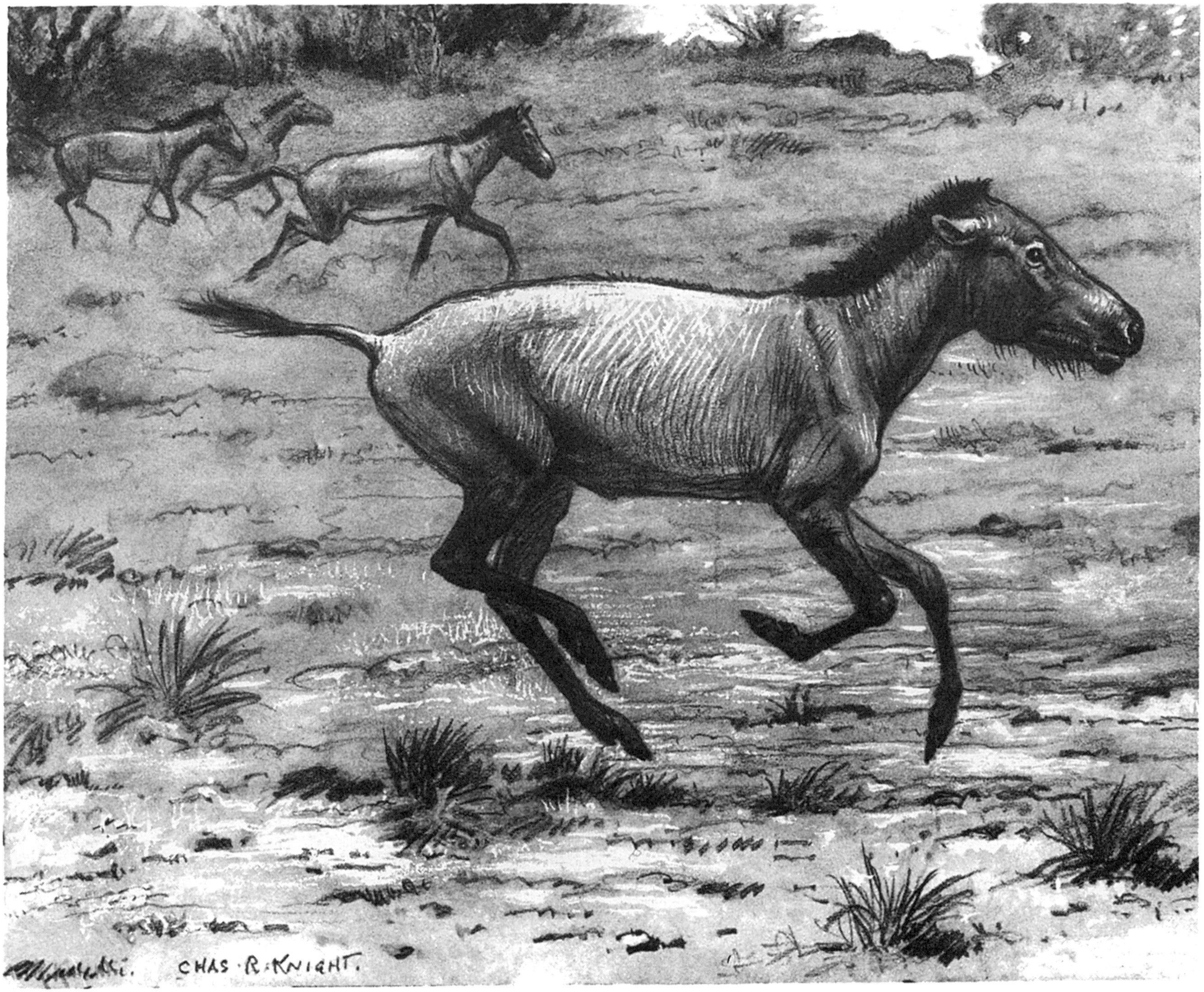

Courtesy American Museum of Natural History

Horses

When King Richard the Third of England cried frantically, "A horse! A horse! My kingdom for a horse!" he demonstrated dramatically the value which all civilized peoples had for ages placed upon this magnificent animal. The Stone Age Man, in still earlier times, valued the horse in a more material way, as he killed and ate it whenever opportunity offered. Evidence of this is found in the huge bone piles discovered in certain parts of Europe. Among these remains all leg bones have been cracked open to obtain the nourishing marrow, a custom which seems to have been universal among primitive races. In a later period, both the Egyptians and the Assyrians made great use of the powerful and sagacious creature, as their wall paintings and bas reliefs depict many battle and hunting scenes, in which they are shown attached in pairs to chariots or with soldiers perched upon their backs. For long periods the United States was the home of countless thousands of wild horses of many species and sizes, some quite small, others as large or even larger than modern varieties. For one reason or another these great herds had all vanished by the time the Indian tribes had spread over the country. In fact, when the first Spanish conquerors landed on our shores, bringing with them their war horses, the Redskins were overawed at the size of the creatures, and thought they were very large deer. However, they soon learned to appreciate the carrying powers of these horses, and many Plains tribes became dependent upon their powerful, wiry steeds, on whose backs they traveled the vast open spaces of our western country. Later, when our own pioneers went westward in search of gold and new homes, they rode and drove horses as a matter of course, and today, cowboys would find themselves helpless without their trained cow-ponies. The species shown, Przewalski's Horse, painted from life, is the only truly wild horse in the world today. It has never been domesticated, never ridden, and is an inhabitant of the Turkish Steppes. The cave drawings of Europe show an animal very like the Przewalski type—short, with an upright mane, no forelock, sparsely-haired tail, and a stocky pony-like body. The Przewalski has a dun-colored body; the tail and mane and legs are dark brown.

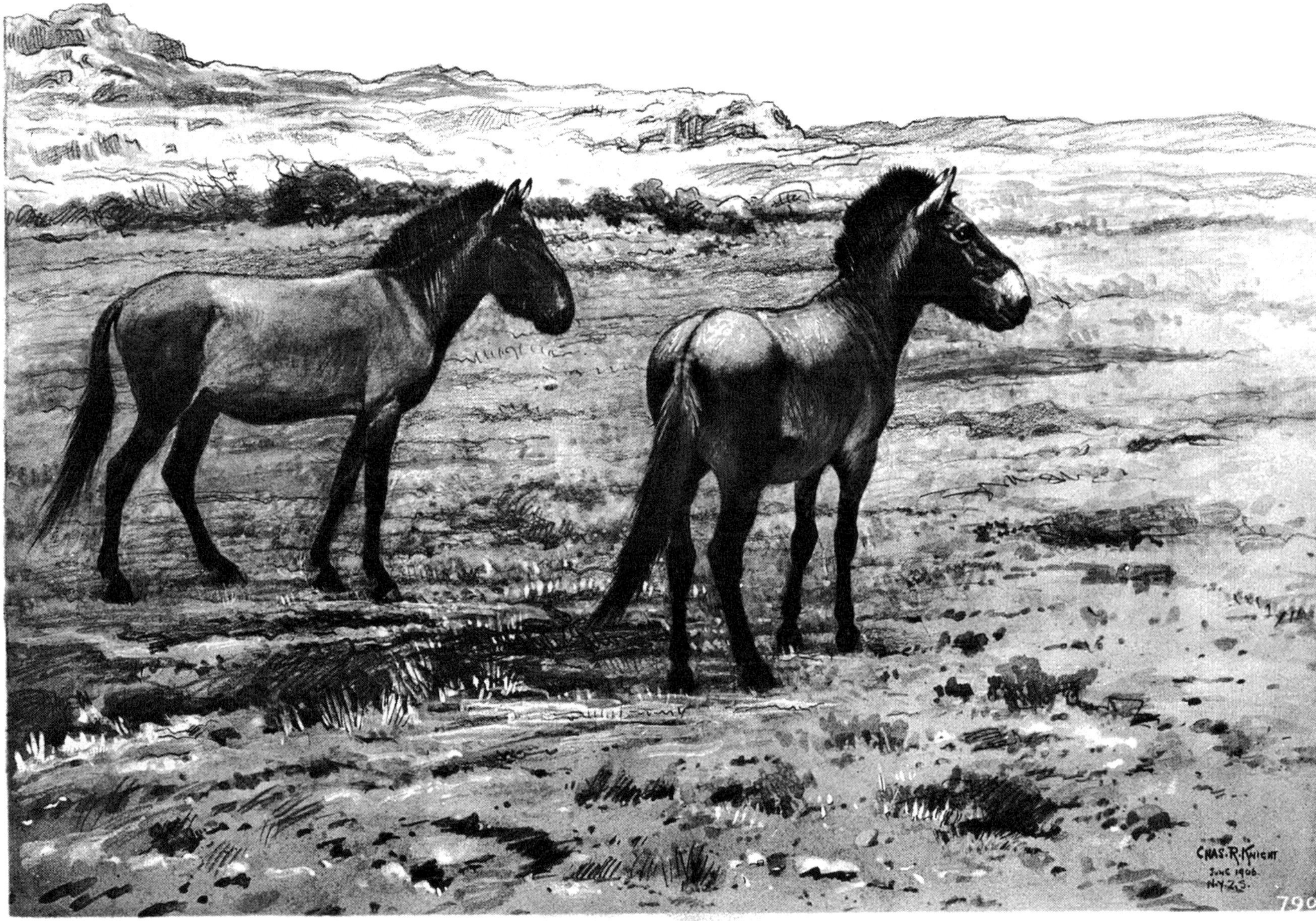

Courtesy American Museum of Natural History

Titanothere—Oligocene Period

The name looks formidable but translated it simply means "big animal." He certainly is all of that, for some species are seven feet at the shoulder, enormously heavy, with a big, broad head and two horny projections at the end of the nose. These horns, placed across the nose, not parallel to it, are not true horns, but are bony growths from the skull, covered with a hardened skin. They must have been of some use to the animal, though it is difficult to say in what way. The big Titanothere is just about the last of a tremendous race of creatures that roamed over our country, and distant Mongolia, in one of the earlv eras of Mammal life. At first, they were small with variously shaped heads and bodies, and were separated into various species. Though not related to rhinoceroses, they must have lived in the same way — roaming about over vast stretches of open country, bathing in streams, and feeding on the abundant plant life of the region. Though the flesh eaters of the time were not particularly large or formidable, they would destroy a straying calf if they could attack it while away from the rest of the herd, much as our modern wolves will kill a buffalo calf today. Though huge, Titanothere was woefully short on brains, a fact which may have had something to do with his final destruction. This cannot be proved, however, and one wonders what set of circumstances, pestilence, or catastrophe could have eventually destroyed a race of animals which appeared to be so strong, so vigorous, and so able to look out for themselves. We do know, however, that in the time of the last Titanotheres in our country, violent volcanic upheavals, with consequent heavy ash and dust falls, were very prevalent, and much food and water was covered and polluted by the acid fumes and lava from the great volcanoes. Whole regions could thus have been devastated until finally the harassed brutes simply lay down and gave up the struggle forever. Today, we find their massive bones under ash and dust deposits, mute evidences of their losing struggle with the elements.

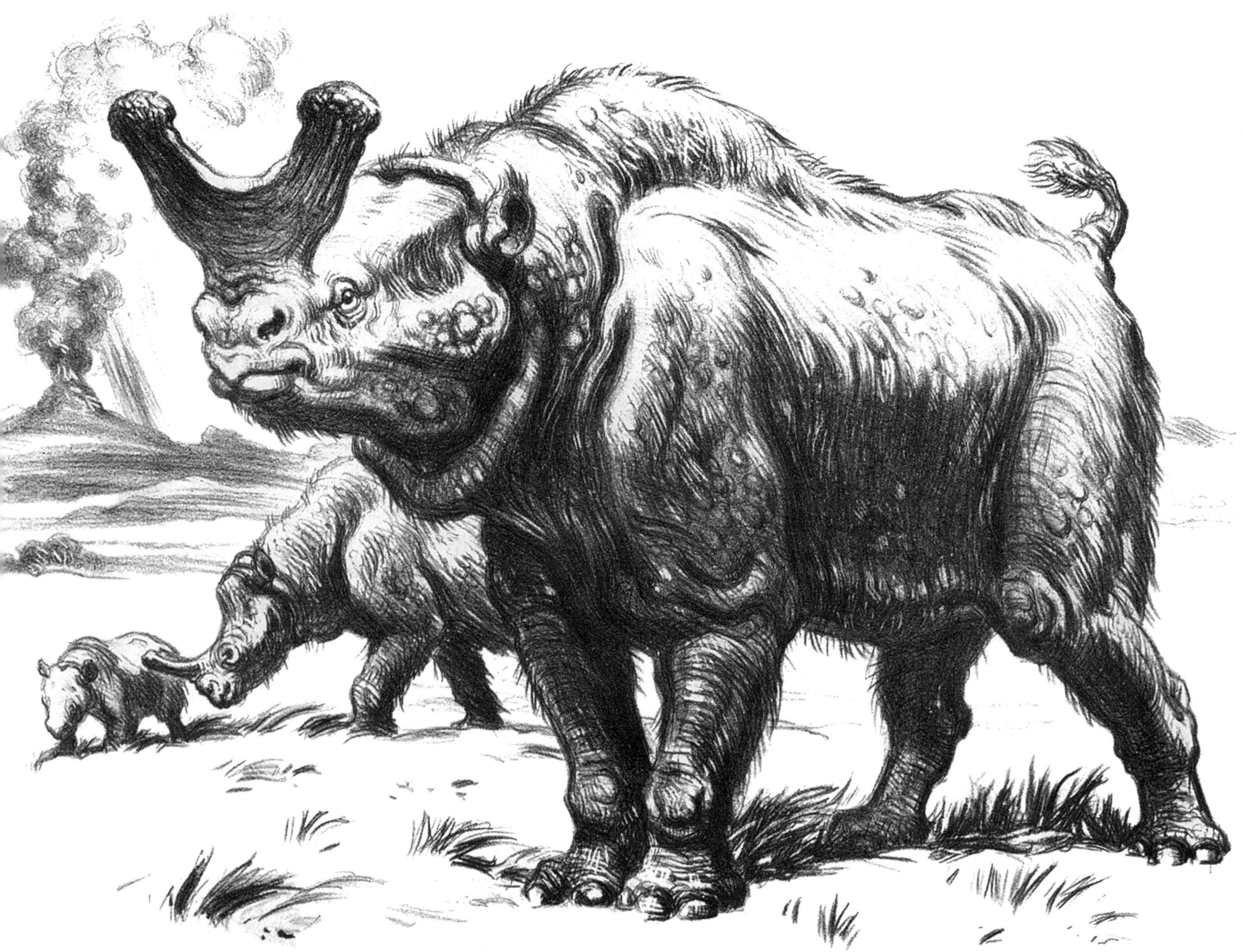

Rhinoceroses

Many years ago the little zoo in Central Park, New York, possessed a fine female specimen of the African black rhinoceros. This singular creature carried two long horns, one behind the other on top of her nose. Armed in this manner she was a dangerous animal. Cranky, sullen, and suspicious, her keepers were very much afraid of her. She could move very quickly for so large an animal. In walking or trotting she had a peculiar springy gait, lifting her feet high at each step. In appearance she was long and rangy, covered with a thick, pinkish, hairless skin. She held her head and neck in a horizontal position except when feeding. The pointed, prehensile lip was very flexible. When presented with a morsel of food she would raise the tip of the lip straight up in what she no doubt considered a begging attitude. She was a strange beast, a kind of leftover from a far-off day.

Still another species, the African square-lipped rhinoceros, together with three Asiatic types—among them the great single-horned Indian—comprise all the recent Rhinoceros family. The Chinese at one time used the upturned horns of the Indian variety as libation cups, for they were supposed to ward off the evil eye. These cups are works of art, beautifully carved and mounted. The horns of the rhinoceros are unique in that they are set upon the tip of the nose and connected with it only by bands of toughened skin. They are solid throughout, not growing over a bony core, as do the horns of all cattle, sheep, goats and antelope. In captivity rhinoceroses have been known to pull these off, but the wound soon heals. Vast numbers of many species of fossil rhinoceroses have lived in this country in the days gone by. They were mostly small, fast-moving animals with either very small horns, or none at all. Hyracodon is one of these early types and you can see in the picture that he was a swift, light-footed, running animal, not so very unlike the horses of his time, though with thicker legs and heavier feet. A gigantic, hornless type (Baluchitherium), the largest land mammal, standing at least sixteen feet at the top of the back, has been found in Baluchistan and Mongolia.

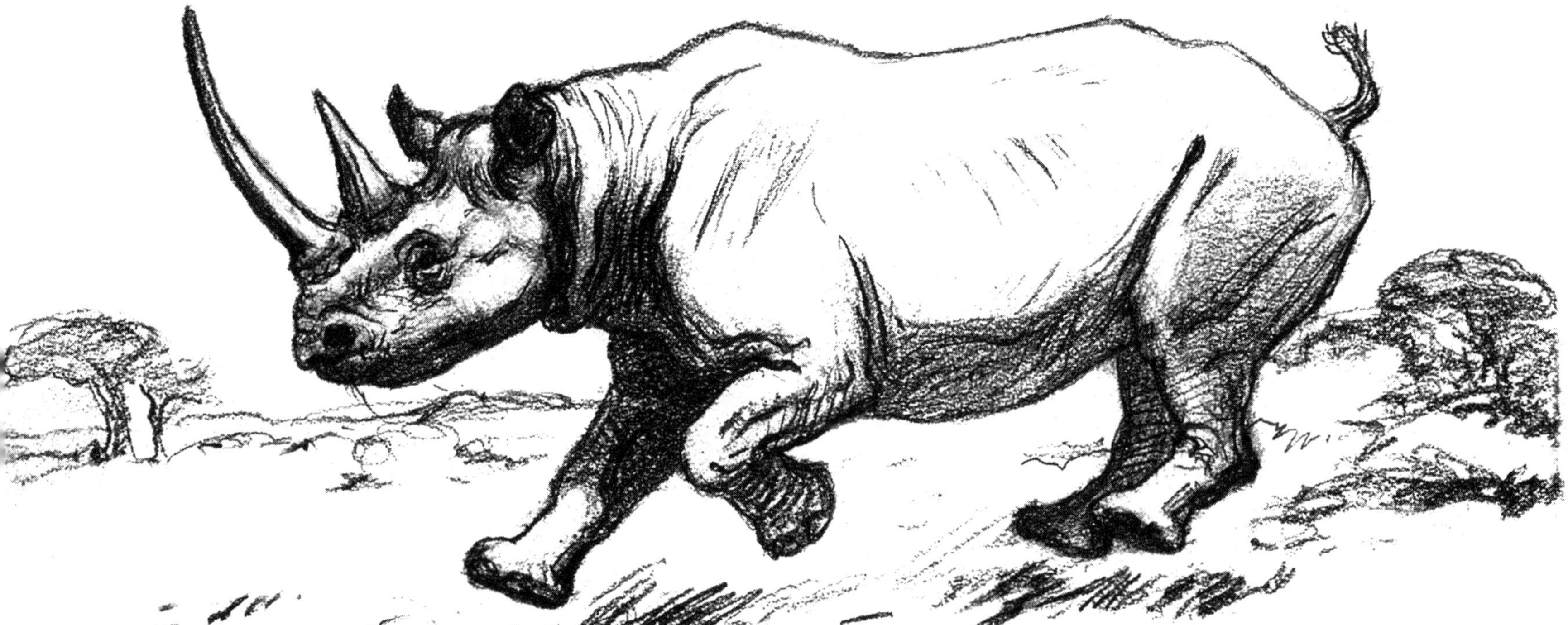

Black Rhinoceros

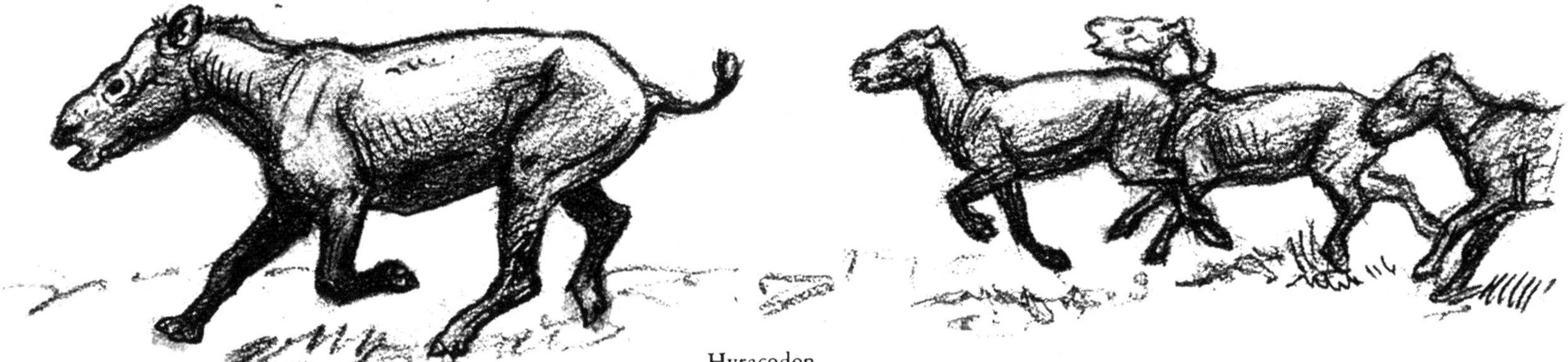

Hyracodon

Courtesy American Museum of Natural History

Woolly Rhinoceros and Man

There is a terrific rush through the gale-swept snow, a fierce upward toss of the great horned head, and the life of our early hunter is snuffed out forever. Later his friends will carry what is left of him back to their shelters, but this will not prevent them from trying once again to kill the big woolly rhinoceros, for they are hungry men and hunger is a terrible driver. Even though the massive, reddish-haired beast with the two long, pointed horns on his forehead may escape them, he will finally succumb to the rigors of an Arctic winter. Thousands of years later we shall again see his shaggy, dark form, quite perfect in its preservation, lifted from some melted mass of tangled grass and bushes, and carried to a scientific museum, there to be studied at leisure. Why such an active and powerful creature as the woolly rhinoceros should ever have elected to stay in the North during the winter is a question difficult to answer. It could easily have gone south before the icy blasts became too vigorous, and lived in a warmer climate. It probably was a browser, eating leaves and buds and tender plant shoots, as do several of its modern relatives. The men of the time who were hunters and who later became agriculturalists, must often have faced starvation in the grim climate of the period. In summer they could and did find game to kill—horses, wild boars and deer—but snow in winter was most difficult to overcome, and it is surprising that they managed to survive at all. It speaks well for man's resourcefulness, sagacity and courage, that without any better weapons than stone axes, spears, and bows and arrows, they were able to provide enough food for themselves and their families. Long before this time man had discovered the use of fire with which to cook his food, keep warm, and scare away wild animals. He still lived in caves or rock shelters. On their walls, drawings of wild animals and of a few human beings can still be seen. Some of the pictures are good, others are not so well done, but they are an attempt to portray the life which surrounded these early men, and which their descendants in later ages were to dominate so completely.

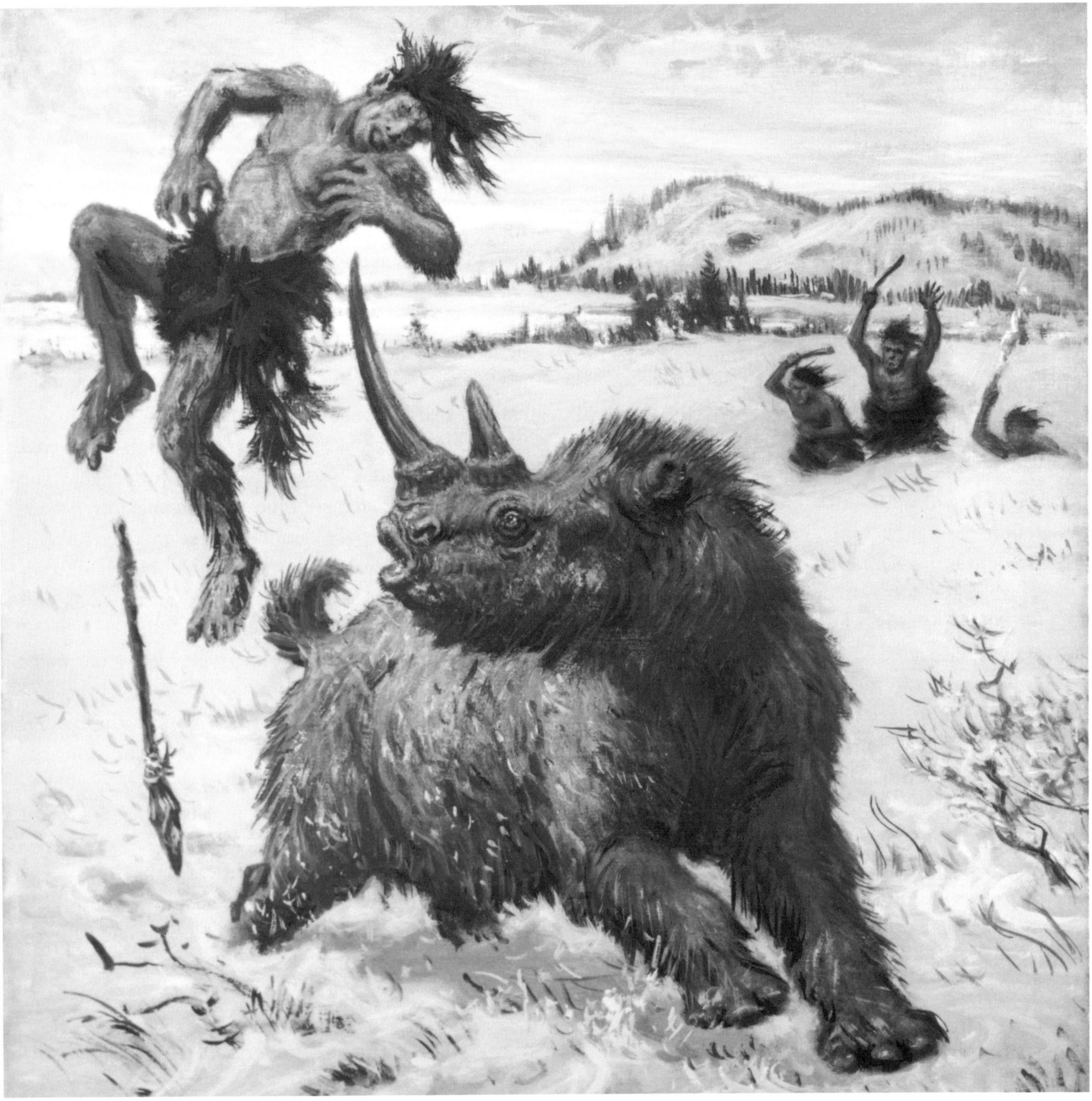

Courtesy American Weekly Magazine

Deer

The timid, gentle deer has long appealed to our imagination as a being of ineffable sweetness, but while this characteristic may apply at certain seasons of the year, it certainly does not hold good for the mating period in the fall. When the horns of the buck attain their full growth, the animal, for the time being, is a decidedly dangerous brute—especially in captivity where he has lost his fear of man. He fights desperately with other bucks of his own species, and two individuals will quite often lock their horns so firmly that they cannot be torn apart. When this happens both animals die a miserable death. The antlers of a deer—shed annually—are bony growths from the skull. They start as the old horns fall off, and passing through a soft velvet-covered stage, they reach their maximum shape and power in the fall and winter months. With these weapons perfected, the deer feels able to conquer his world, and to beat off attacks of wild animals. Small deer, of various forms, figure quite early in the Age of Mammals, but the race attained its finest development at a much later period. The Irish Elk (Megaceros), with superb horns, some ten feet across at the tip, is one of these more recent types. He was not an elk but an ancestral fallow deer (the pretty little spotted animal one sees now living in European parks), and his fossil remains are quite frequently found buried in the Irish bogs. Specimens of the male animal are more commonly found than those of the opposite and hornless sex, owing probably to the fact that the great weight of the male's horns had a tendency to unbalance the animal when drinking. Cervalces, a curious moose-like animal, is not a real moose but a prehistoric species, whose skeleton was discovered all complete in a Jersey bog some years ago. The moose himself, that six-foot mass of ungainly power, is in some ways the grandest deer that ever lived. Immensely tall, and very short-bodied, his great head, widespread palmated antlers, extraordinary nose, and long mule-like ears, combine to make him king of all antlered creatures. He swims well, feeds on water lily roots and the bark of trees, and can travel through heavy forests at a speed which astonishes the hunter.

Moose do not thrive in captivity, and for that reason one never sees them in good condition unless in a state of nature. The most superb examples are found in the forests of Alaska. The Eastern moose, those found in Maine and New Brunswick, are also of very great size and majestic mien. The European species, known in Europe as the elk, is very similar to ours, though its antlers are smaller.

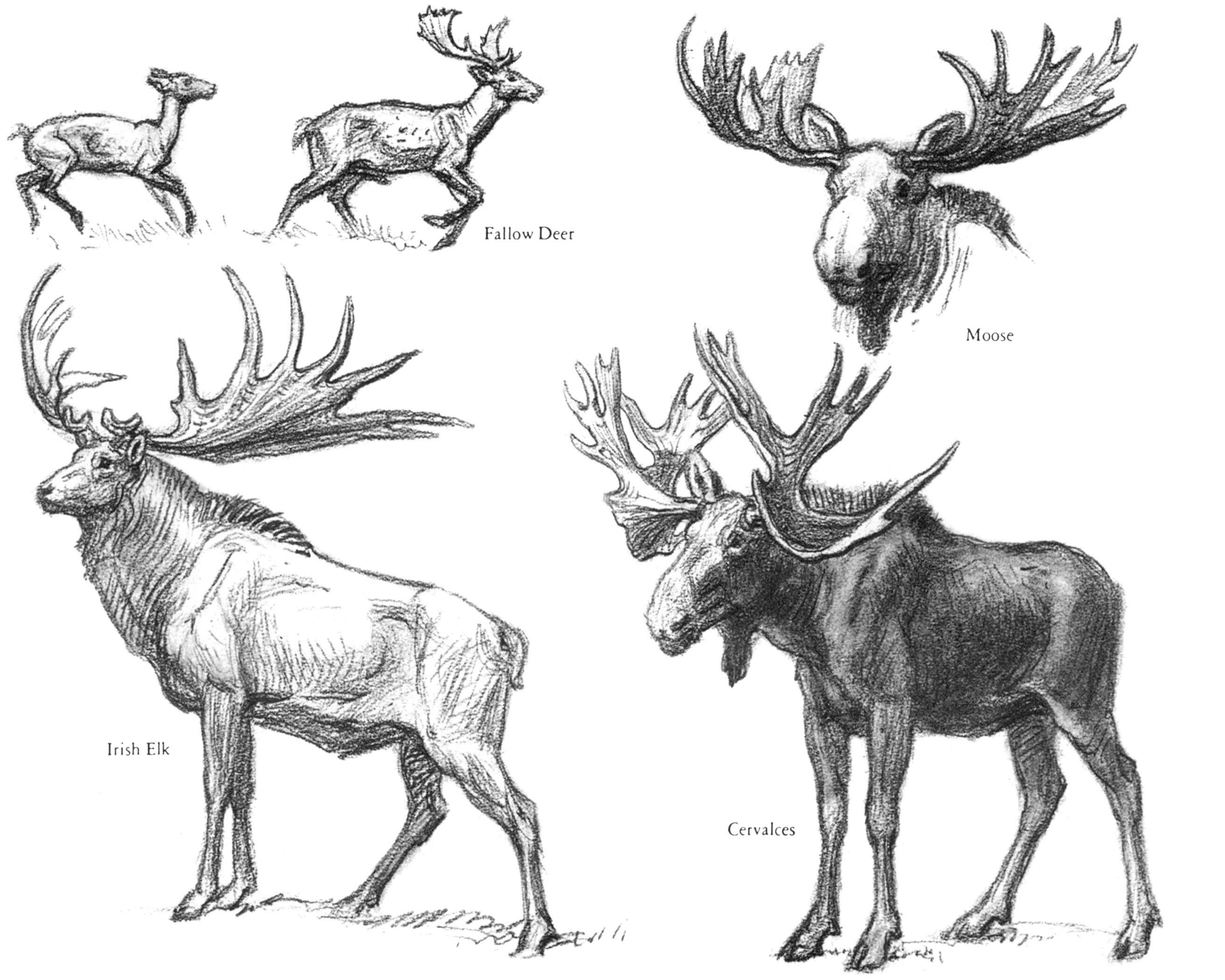
Fallow Deer
Moose
Irish Elk
Cervalces

Sloths

Today the sloths are a small and insignificant family. There are only two small species which creep around on the underside of branches of trees in South American forests. But there was a time when the great ancestors of these little nonentities roamed over a considerable part of both North and South America. Megatherium, largest of this fabulous group, was an immensely powerful, slow-moving beast, with huge forelimbs armed with stout claws. Its short, stocky hind legs were also provided with big, digging toes. Six feet across the hips, he attained a height of fifteen feet or more when in a sitting position. He was indeed a giant in every way—except in the matter of his mentality which was of a low order. No doubt he dug for tubers and roots with huge front claws and in spite of his formidable appearance was undoubtedly a peacefully-inclined monster. He would defend himself only when attacked by some great flesh eater like the predatory sabre-toothed tiger, which lived in the same region. A long prehensile tongue protruded at times from his tube-like mouth, and curling round some bit of succulent greenery would pull the leaves into his capacious jaws. Fortunately, we know much about the life appearance of this great creature—not only the color and quality of the hair of an almost similar species, but even the texture of his skin, and the make-up of his complete skeleton. Some years ago, in a dry cave in Patagonia, explorers discovered all this evidence and much more besides, pointing to the probable enforced occupation of the cave by a great Mylodon, no doubt held captive by Indians. A large amount of grass lay about, and the bones with flesh removed, but covered with particles of dried blood, were fresh in appearance. If the great sloth is queer-looking, what may be said of the heavily-armored animal at his feet? Powerful muscles and a heavy bony frame lie under that great carapace with its war-club tail. We call this worthy a Glyptodon, a relative of the sloth and of the modern armadillo. He couldn't roll up as does the armadillo, but the movement was not necessary because of his size and strength. Harmless, but thoroughly armed for defence, these extraordinary creatures finally disappeared, until today only the little tree-hanging sloths and a few species of armadillos are left to represent this once powerful race.

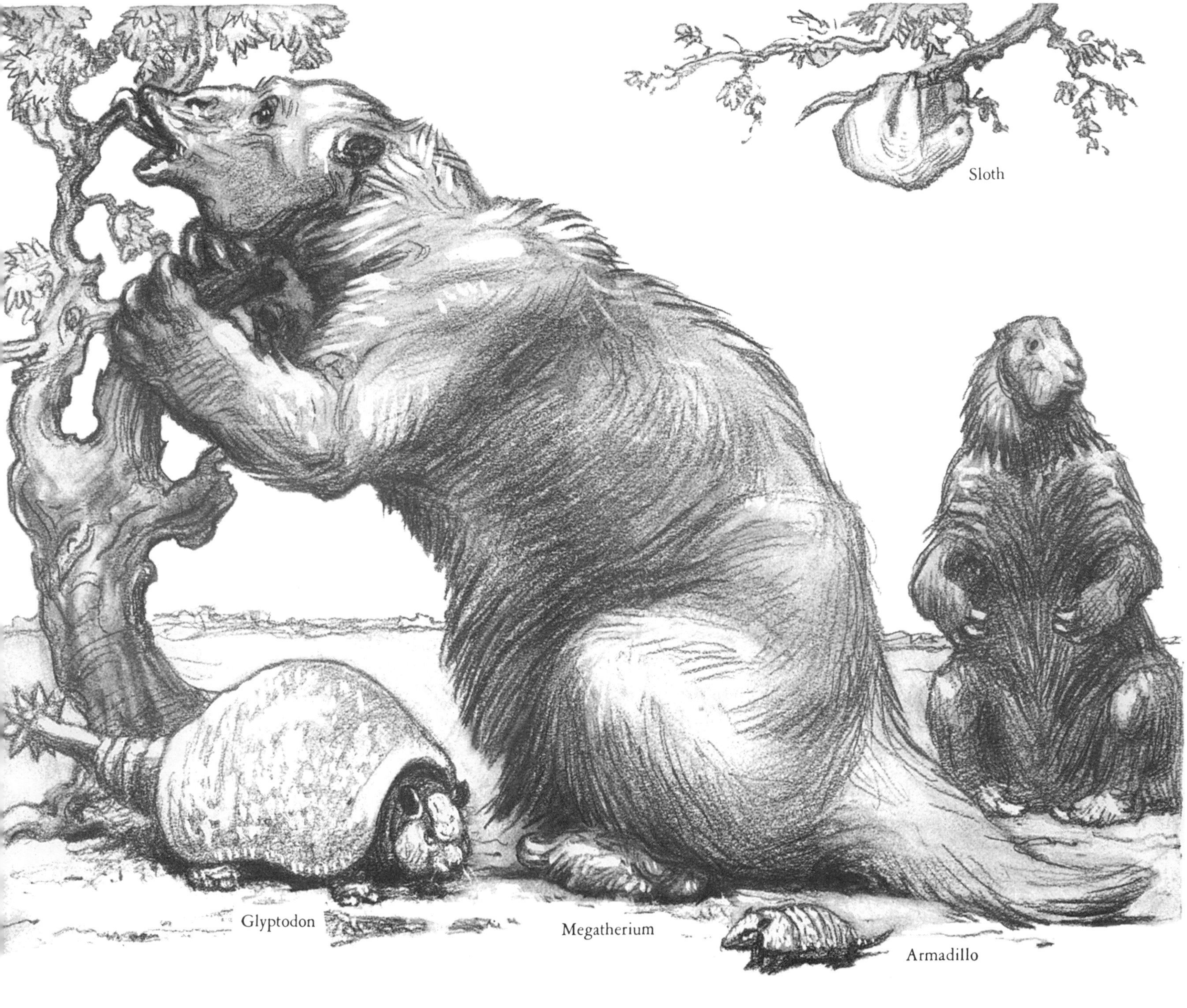
Sloth
Glyptodon
Megatherium
Armadillo

Bison and Wolves

We don't usually think of Florida as the home of the bison, yet we know from the remains of the creatures that have been found that they did once exist there. Food must have been more abundant then, as Florida today has difficulty supporting a small number of undernourished cattle, and nothing larger than a deer among the wild stock. Besides the bison, there were horses, ground sloths, mammoths—all great consumers of green things—and, of course, there were flesh eaters to prey upon them. The bison bull in the picture is being attacked by large timber wolves, terrible biters, and able with luck to kill the heavy brute by a slashing bite across the hock and bring him down with a crash, thus rendering him perfectly helpless. You will notice not only the vast bulk of the bison but also the great horns of this animal—so much longer than those of our present species. It seems almost incredible when we read that up to seventy or eighty years ago there were estimated to be some fifty million of these splendid creatures roaming our western Plains country from Canada to Texas. It was the white man who encompassed their near-extinction in the early '80's of the last century, and it was only by the strenuous efforts of a few determined men, that the pitiful remnants of that once vast company were saved for posterity. Today, a goodly number thrive and multiply in the wild country of Canada—great, heavily-furred animals, inured to the bitter cold and snow of their enforced environment. As the bison are our grandest and largest mammals, we should be proud to preserve these last stragglers of a once mighty race. In Europe, also, until 1917, there were still a small number of the great European bison or wisent, a few of these magnificent creatures running practically wild in one of the great parks belonging to a Russian nobleman. But during the Russian Revolution nearly all of these unfortunate creatures were destroyed. However, a few were left in German zoological gardens though these, too, may now be gone. It seems a pity that these magnificent beasts should have been killed and the race practically exterminated.

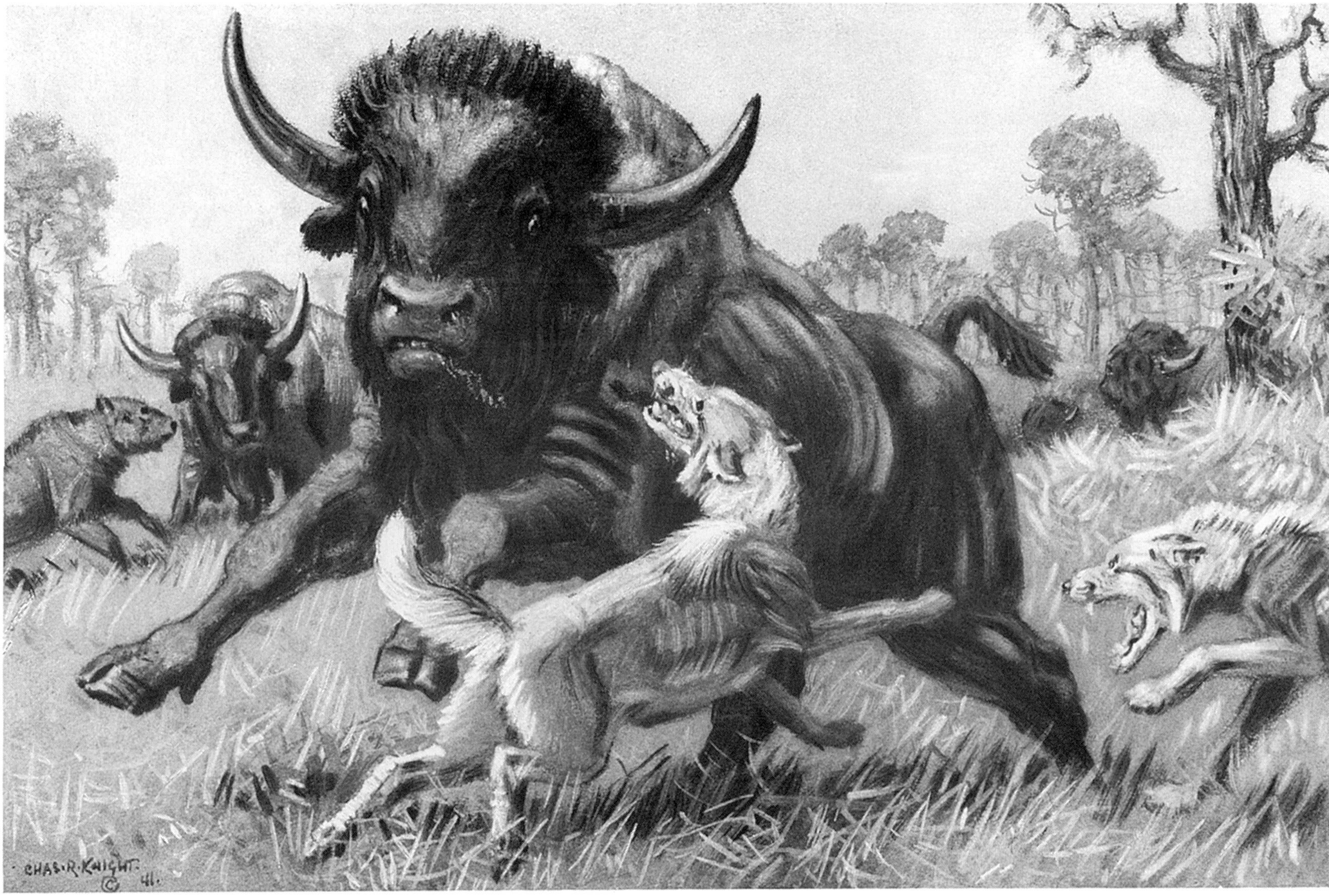
CHAS·R·KNIGHT·
© 41·

Gorilla

"Ze great Beess—he stan' up—he beat his bres' wiz hees beeg feests—he roar—I shoot—CRASH! He fall dead at my feet—I have keel my gorilla." In these words, Paul du Chaillu, noted African explorer, described to me his first gorilla hunt. In his day many things which he told were discredited, but since then, other hunters have verified most of his statements regarding the size, power, and forbidding aspect of the great primate. The stiffly-squared figure in the drawing is "Bushman"—four-hundred-pound monster at the Lincoln Park Zoo in Chicago. The powerful animal is in a decidedly captious frame of mind and his intentions toward me are hostile to a degree. If I am foolish enough to stroll behind his cage, a shower of fruit, vegetables, eggs, et cetera, will be hurled in my direction, all intended to prove that Bushman doesn't like my presence and is taking this insulting way of showing it. On my last visit to the Chicago Zoo, Bushman was on a rampage, not even the keepers daring to pass behind his cage; yet the cause for this show of temper is not easy to ascertain. Perhaps his age—he is now fourteen—may have something to do with it. On seeing me he would rise, towering above me, beating his breast and rushing across the cage to strike me down with his great hands. At such times he was, in reality, "Ze great Beess," so dramatically described by Du Chaillu. Physically, male gorillas are huge, six feet tall, and immensely broad and muscular. There are two varieties, one found in the Congo, the other on the west coast of Africa.

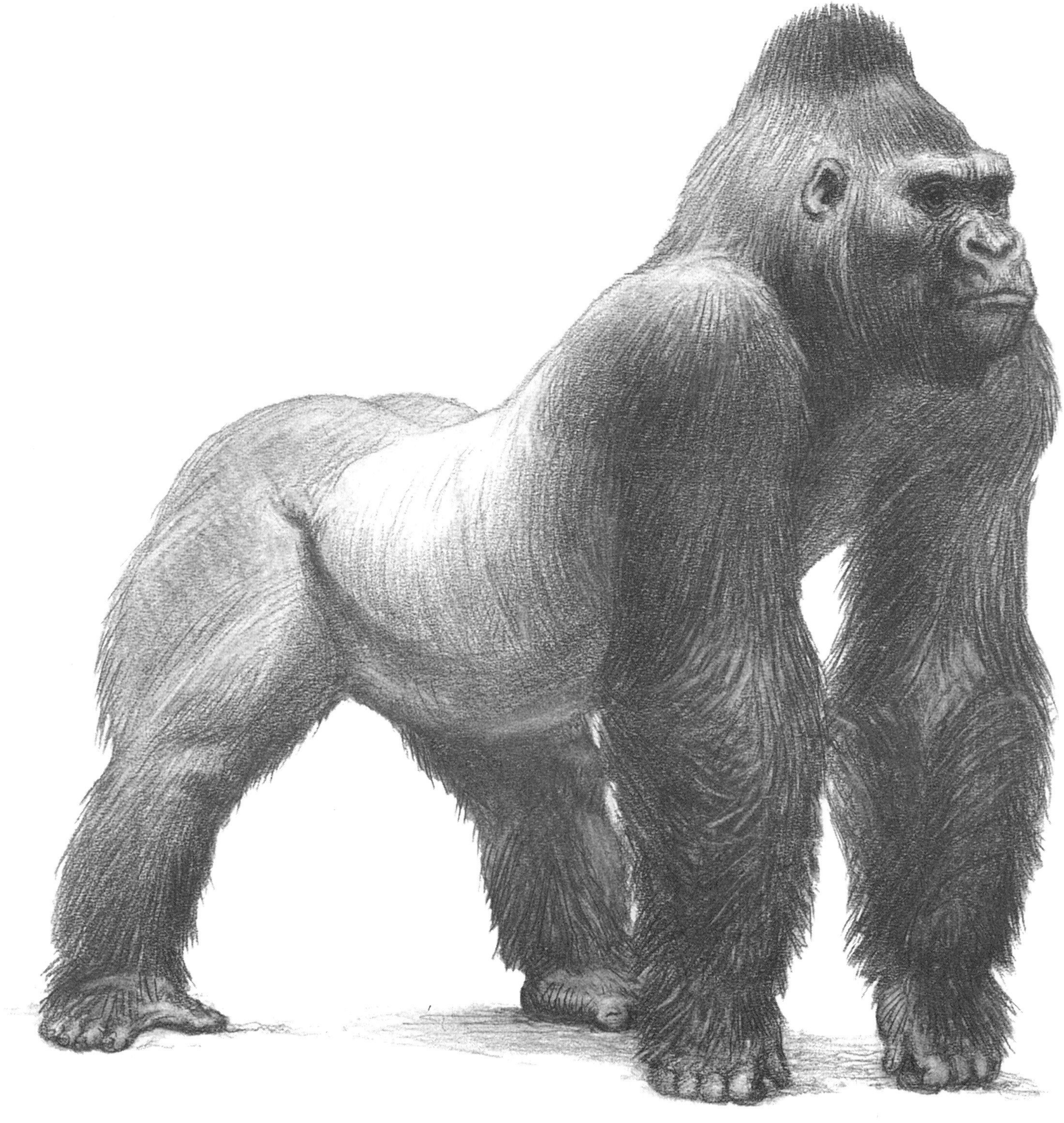

Cave Bear and Men

It must have taken nerve, strength, and skill, to fight with a monster so powerful and so fast as the great cave bear. Like modern bears, the huge brute was no doubt an expert boxer, hurling right and left any weapon aimed at his enormous head, and dealing terrific blows with his long chisel-like claws. The vast creature must have been a source of dread to our prehistoric forefathers, disputing cave habitation with them, and stealing from their food caches whenever possible. Undoubtedly, these great animals would make short work of any man if they could once strike or bite him. But in the main, they were not so much of a menace as the cave lions, hyenas, and huge wolves, which roamed the countryside in packs. All these fierce, predaceous animals would regard a defenseless human as a mere tidbit to be gobbled up in passing. There is no doubt that man often smoked the cave bear out of his den, and when the beast appeared, groggy from the fumes, struck him over the head with a stone or wooden implement. In fact, there is somewhere in Europe a cave bear skull with a stone axe still sticking in it. They were very common animals in their day and great numbers of their bones have been taken from the European caves. While they were really huge, with immense heads, their legs were short so that they could never have had the bulk of our great Kodiac bear. Bears, as a family, have a lengthy history. To judge from their teeth, their food has always been of a very general nature, almost anything from very large to absurdly small things being accepted as sustenance. Our own great grizzly bears will eat ants, slugs, ground squirrels, fruits, nuts, honey (bees included), fish, horses, deer, pigs, and many other things. Truly, they do not lack vitamins in their diet. To most people bears are very amusing animals, particularly when young, but they strike a sinister note with their vast bulk, peculiar, leering mouth, tiny eyes, the impression of gigantic power exhibited in the huge muscular frame, their awkward gait, and their man-like attitude in the erect position.

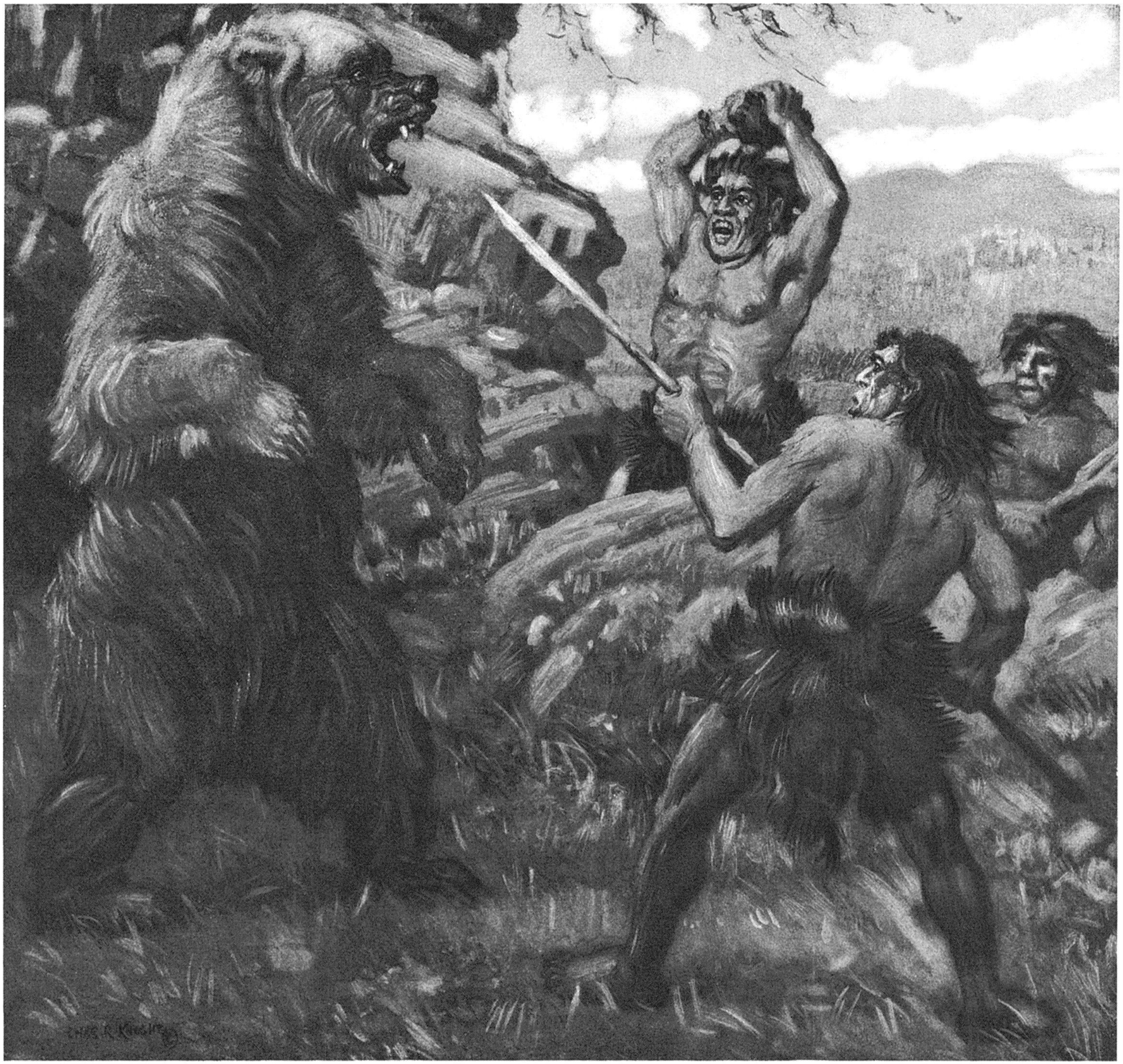

Courtesy American Weekly Magazine

CHARLES ROBERT KNIGHT (1874–1953) was a distinguished paleolontological artist. His work remains on display at the American Museum of Natural History, the Field Museum, and elsewhere, where it continues to inspire viewers with its wonderfully detailed reconstructions of lost worlds and long extinct animals.